Astrophysics for the Curious

Decoding the Universe, Black Holes, and the Big Bang

Nico Quinn

© Copyright 2024 by Nico Quinn
All Rights Reserved

This document is geared towards providing exact and reliable information with regards to the topic and issue covered. The publication is sold with the idea that the publisher is not required to render accounting, officially permitted, or otherwise, qualified services. If advice is necessary, legal or professional, a practiced individual in the profession should be ordered. From a Declaration of Principles which was accepted and approved equally by a Committee of the American Bar Association and a Committee of Publishers and Associations. In no way is it legal to reproduce, duplicate, or transmit any part of this document in either electronic means or in printed format. Recording of this publication is strictly prohibited and any storage of this document is not allowed unless with written permission from the publisher. All rights reserved. The information provided herein is stated to be truthful and consistent, in that any liability, in terms of inattention or otherwise, by any usage or abuse of any policies, processes, or directions contained within is the solitary and after responsibility of the recipient reader. Under no circumstances will any legal responsibility or blame be held against the publisher for any reparation, damages, or monetary loss due to the information herein, either directly or indirectly. Respective authors own all copyrights not held by the publisher. The information herein is offered for informational purposes solely, and is universal as so.

The presentation of the information is without contract or any type of guarantee assurance. The trademarks that are used are without any consent, and the publication of the trademark is without permission or backing by the trademark owner. All trademarks and brands within this book are for clarifying purposes only and are the owned by the owners themselves, not affiliated with this document.

Table of Contents

Chapter 1 6

Introduction to Astrophysics 6

The Fascination with the Cosmos 6

Historical Perspectives From Ancient Astronomy to Modern Astrophysics 10

Key Concepts and Terminology 15

Tools and Techniques in Astrophysics 19

How to Use This Book 24

Chapter 2 29

The Universe An Overview 29

The Scale of the Universe 29

Observing the Cosmos Telescopes and Space Missions 33

The Electromagnetic Spectrum: Seeing the Invisible 37

The Expanding Universe and Hubble's Law 42

Dark Matter and Dark Energy 46

Chapter 3 51

The Birth of the Universe The Big Bang Theory 51

The Origins of the Big Bang Theory 51

Evidence Supporting the Big Bang 55

Cosmic Microwave Background Radiation 59

Nucleosynthesis Formation of the First Elements 64

The Evolution of the Universe Post-Big Bang 68

Chapter 4 73

Galaxies Islands in the Universe 73

Types of Galaxies Spiral, Elliptical, and Irregular .73

The Milky Way Our Galactic Home.......................77

Galaxy Formation and Evolution 81

Interactions and Collisions Between Galaxies 85

Active Galactic Nuclei and Quasars....................... 90

Chapter 5 95

Stars The Building Blocks of Galaxies...................... 95

Star Formation From Nebulae to Protostars 95

The Life Cycle of Stars Main Sequence to Supernova
.... 99

Types of Stars Dwarfs, Giants, and Neutron Stars
....103

Stellar Nucleosynthesis Birth of Elements...........107

The Hertzsprung-Russell Diagram 111

Chapter 1

Introduction to Astrophysics

The Fascination with the Cosmos

Since the dawn of humanity, the night sky has captivated our imaginations. The vast expanse of twinkling stars, the mysterious glow of distant planets, and the occasional streak of a shooting star have all fueled a collective curiosity that spans cultures and generations. This fascination with the cosmos is as old as civilization itself, intertwining with our myths, our science, and our quest for understanding our place in the universe.

Ancient civilizations were some of the first to document their observations of the cosmos. The Babylonians, Egyptians, and Mayans, among others, meticulously recorded the movements of celestial bodies, creating early astronomical charts and calendars. These records were not just for scholarly pursuit; they were deeply integrated into the societies' religious and daily lives. The alignment of the stars and planets was believed to influence everything from agricultural cycles to the fates of empires. The ancient Greeks, particularly through the works of philosophers like Ptolemy and Aristotle, laid the groundwork for a more systematic study of the heavens, developing early theories about the structure of the cosmos that would influence scientific thought for centuries.

The Renaissance period marked a significant shift in our understanding of the cosmos. With the advent of the telescope, Galileo Galilei's groundbreaking observations challenged the long-held geocentric model of the universe, which placed Earth at its center. Instead, Galileo's work supported the heliocentric model proposed by Copernicus, positioning the Sun at the center of our solar system. This paradigm shift was not just a scientific revolution but a profound change in how humanity perceived its place in the cosmos. It highlighted the dynamic and ever-evolving nature of our understanding of the universe.

The 20th century brought about unprecedented advancements in our exploration of space. The launch of the Hubble Space Telescope in 1990 provided humanity with some of the most detailed and awe-inspiring images of the universe ever captured. Hubble's observations have given us glimpses into the distant past, revealing galaxies billions of light-years away and helping to refine our understanding of cosmic phenomena such as black holes and dark matter. These images and data have not only advanced our scientific knowledge but also deepened our sense of wonder about the universe's vastness and complexity.

The moon landing in 1969 was another pivotal moment in our cosmic journey. Neil Armstrong's famous words, "That's one small step for man, one giant leap for mankind," echoed the sentiment of an entire generation witnessing the realization of a dream that had seemed fantastical just decades earlier. The Apollo missions demonstrated human ingenuity and the potential for future exploration, inspiring countless individuals to look to the stars with a renewed sense of possibility.

Beyond the scientific and technological milestones, the cosmos has always been a rich source of inspiration for art, literature, and philosophy. The Romantic poets, such as William Wordsworth and John Keats, often drew on the beauty and mystery of the night sky to convey deeper emotional and existential themes. Vincent van Gogh's iconic painting "Starry Night" captures the swirling, almost dream-like quality of the night sky, reflecting the artist's own turbulent inner world. In literature, the cosmos serves as both a setting and a metaphor for the human experience, from the epic voyages in Homer's "The Odyssey" to the speculative realms of science fiction.

Science fiction, in particular, has played a crucial role in shaping public interest and understanding of space. Works by authors like Isaac Asimov, Arthur C. Clarke, and Philip K. Dick explore the possibilities of interstellar travel, encounters with extraterrestrial life, and the ethical implications of advanced technologies. These stories, while often fantastical, are grounded in scientific principles and provoke thoughtful consideration of our future in space. They challenge us to imagine the unknown and to confront the potential consequences of our actions as we venture further into the cosmos.

The fascination with the cosmos is also deeply personal. For many, gazing up at the night sky evokes a sense of awe and humility, a reminder of the vastness beyond our everyday concerns. It connects us to something greater than ourselves, prompting existential questions about life, the universe, and everything in between. This sense of wonder is not limited to professional astronomers or scientists; it is a universal experience that transcends cultural and geographical boundaries.

In recent years, the rise of amateur astronomy has made the cosmos more accessible to the public. Affordable telescopes, online stargazing communities, and citizen science projects allow individuals to participate in astronomical research and observation. Events like meteor showers, eclipses, and planetary alignments draw people together to share the experience of witnessing celestial phenomena. This democratization of astronomy fosters a collective appreciation for the universe and encourages a lifelong pursuit of knowledge and discovery.

The future of our fascination with the cosmos holds exciting possibilities. Space agencies like NASA and ESA, along with private companies like SpaceX, are planning ambitious missions to Mars, asteroids, and beyond. These missions aim not only to explore new frontiers but also to address fundamental questions about the origins of life and the potential for habitable environments elsewhere in the universe. The development of space tourism promises to make space travel a reality for more people, offering the unparalleled experience of seeing Earth from orbit and fostering a deeper connection to our planet.

Moreover, advances in technology and data analysis are transforming our ability to study the cosmos. Artificial intelligence and machine learning algorithms are being used to sift through vast amounts of astronomical data, identifying patterns and anomalies that would be impossible for humans to detect alone. These tools are helping to uncover new exoplanets, map the distribution of dark matter, and study the cosmic microwave background radiation, bringing us closer to understanding the fundamental nature of the universe.

Historical Perspectives From Ancient Astronomy to Modern Astrophysics

Ancient civilizations gazed at the night sky with a mix of wonder, reverence, and curiosity. The earliest astronomers were not just scientists but also priests, poets, and philosophers. They observed the heavens to understand the divine, to mark the passage of time, and to navigate their journeys. The Babylonians, for instance, meticulously recorded celestial events on clay tablets, creating some of the first known star catalogs. These early astronomers noticed patterns in the movements of the sun, moon, and stars, leading to the development of the zodiac and the prediction of eclipses.

In Egypt, the alignment of the pyramids and temples with certain stars and constellations reflected a deep connection between the heavens and earthly life. The Great Pyramid of Giza, for instance, is aligned with the North Star, symbolizing the pharaoh's journey to the afterlife. The ancient Greeks made significant strides in theoretical astronomy. Pythagoras proposed that the Earth was spherical, a revolutionary idea at the time. Later, Aristotle provided a more detailed argument for a spherical Earth, based on observations of the Earth's shadow on the moon during a lunar eclipse.

Aristotle's influence was profound, but it was Claudius Ptolemy who provided the most comprehensive model of the cosmos in antiquity. Ptolemy's geocentric model, described in the "Almagest," placed Earth at the center of the universe, with the sun, moon, planets, and stars revolving around it in complex orbits. This model dominated astronomical thinking for over a thousand years. In the Islamic Golden Age, scholars such as Al-Battani and Al-Sufi built upon Greek and Indian astronomy, making precise observations and corrections to earlier models. The translation of their works into Latin during the Middle Ages helped preserve and transmit astronomical knowledge to Europe.

The Renaissance brought a paradigm shift in our understanding of the cosmos. Nicolaus Copernicus, in his seminal work "De revolutionibus orbium coelestium," proposed a heliocentric model, placing the sun at the center of the universe and Earth as one of the planets orbiting it. This idea was radical and met with resistance, but it set the stage for a scientific revolution. Galileo Galilei's use of the telescope provided empirical support for the Copernican model. His observations of the moons of Jupiter, the phases of Venus, and the rugged surface of the moon challenged the Aristotelian view of a perfect, unchanging heavens. Galileo's advocacy for the heliocentric model led to conflict with the Roman Catholic Church, culminating in his famous trial and house arrest.

Johannes Kepler's laws of planetary motion further refined the heliocentric model. By analyzing the meticulous observations of Tycho Brahe, Kepler discovered that planets move in elliptical orbits, not perfect circles as previously thought. This insight provided a more accurate description of planetary motions and laid the groundwork for Isaac Newton's law of universal gravitation. Newton's "Philosophiæ Naturalis Principia Mathematica" unified the heavens and the Earth under a single theory of gravity. Newton showed that the same force that causes an apple to fall to the ground also governs the motion of the planets. This synthesis marked the birth of modern astrophysics, transforming our understanding of the universe into a coherent, predictable system governed by physical laws.

The 19th century saw the rise of observational astronomy. William Herschel's discovery of Uranus expanded the known boundaries of the solar system. His detailed cataloging of nebulae and star clusters revealed the vast and complex structure of our galaxy. The invention of the spectroscope allowed astronomers to analyze the composition of stars, leading to the realization that stars are composed of the same elements found on Earth. The 20th century brought about a revolution in our understanding of the cosmos, beginning with Albert Einstein's theory of general relativity. Einstein's equations predicted the existence of black holes and the expansion of the universe, ideas that were initially met with skepticism but later confirmed by observation.

Edwin Hubble's discovery of the redshift-distance relationship provided the first empirical evidence for the expanding universe. Hubble's observations of distant galaxies showed that the universe was not static but dynamic, constantly stretching and evolving. This led to the development of the Big Bang theory, which posits that the universe began as a singularity and has been expanding ever since. The mid-20th century also saw the dawn of radio astronomy. The detection of radio waves from space revealed previously unseen aspects of the universe, such as pulsars and quasars. These discoveries challenged existing models and opened new avenues of research in high-energy astrophysics.

The launch of space telescopes, such as the Hubble Space Telescope in 1990, provided unprecedented views of the universe. Hubble's deep field images captured thousands of galaxies in a single frame, some of which were billions of light-years away. These observations have helped refine our understanding of the universe's age, structure, and evolution. The discovery of the cosmic microwave background radiation in 1965 provided strong evidence for the Big Bang theory. This faint glow, a remnant of the early universe, has been mapped in exquisite detail by missions such as COBE, WMAP, and Planck. These maps have revealed subtle fluctuations in temperature that correspond to the seeds of galaxies and large-scale structures we see today.

In recent decades, the field of astrophysics has been revolutionized by the discovery of exoplanets. The Kepler Space Telescope has identified thousands of planets orbiting other stars, some of which are located in the habitable zone where liquid water could exist. These discoveries have profound implications for the search for extraterrestrial life and the understanding of planetary systems. The detection of gravitational waves in 2015 by the LIGO and Virgo collaborations opened a new window on the universe. These ripples in spacetime, caused by violent events such as the merging of black holes, provide a new way to observe and study the cosmos. Gravitational wave astronomy is still in its infancy, but it promises to unveil previously hidden aspects of the universe.

Today, the field of astrophysics is marked by international collaboration and the deployment of increasingly sophisticated instruments. The James Webb Space Telescope, scheduled for launch in 2021, promises to peer deeper into the cosmos than ever before, potentially uncovering the first galaxies that formed after the Big Bang. Ground-based observatories, such as the Extremely Large Telescope (ELT) and the Square Kilometre Array (SKA), will complement these efforts, providing detailed observations across a wide range of wavelengths.

As we continue to explore the universe, we are constantly reminded of the interconnectedness of all things. The study of the cosmos is not just about understanding distant stars and galaxies; it is about understanding our own origins and our place in the grand tapestry of existence. From the ancient astronomers who first mapped the stars to the modern astrophysicists probing the depths of space, our journey through the cosmos is a testament to the enduring human spirit of exploration and discovery.

Key Concepts and Terminology

Understanding the key concepts and terminology in any field is essential for building a solid foundation of knowledge, and this is especially true in astronomy and astrophysics. Grasping these fundamental terms will help you navigate the complex and fascinating world of the cosmos with greater ease and clarity.

Astronomy, the study of celestial objects and phenomena, has a rich vocabulary that reflects its long history and diverse areas of focus. One of the most basic units of measurement in astronomy is the astronomical unit (AU), which is defined as the average distance between Earth and the Sun, approximately 93 million miles or 150 million kilometers. This unit is a convenient way to express distances within our solar system. For example, Jupiter is about 5.2 AU from the Sun.

Moving beyond our solar system, distances become vast and are often measured in light-years, the distance light travels in one year, about 5.88 trillion miles or 9.46 trillion kilometers. This concept helps us comprehend the immense scales involved when studying stars, galaxies, and the universe. For instance, the nearest star system, Alpha Centauri, is about 4.37 light-years away from us.

Stars, the most familiar celestial objects, are massive spheres of plasma that emit light and heat due to nuclear fusion occurring in their cores. The Sun, our closest star, is a G-type main-sequence star, commonly referred to as a yellow dwarf. Stars come in various types and sizes, classified by their spectral characteristics and luminosity. The Hertzsprung-

Russell diagram is a key tool for understanding the relationship between a star's brightness, color, and evolutionary stage. Main-sequence stars, like the Sun, form a diagonal band on this diagram, while giants, supergiants, and white dwarfs occupy distinct regions.

The lifecycle of a star is a fascinating journey from birth to death. Stars form from clouds of gas and dust, known as nebulae, under the influence of gravity. As the material collapses, it heats up and forms a protostar. Once nuclear fusion ignites in the core, the star enters the main sequence phase, where it spends most of its life. Eventually, the star exhausts its nuclear fuel, leading to various end states depending on its mass. Low-mass stars become white dwarfs, while massive stars may explode as supernovae, leaving behind neutron stars or black holes.

Planets, the celestial bodies that orbit stars, are classified into different types based on their composition and location within a planetary system. Terrestrial planets, like Earth, are rocky and located close to their parent star. Gas giants, such as Jupiter, are composed mainly of hydrogen and helium and are found further out in the solar system. Ice giants, like Neptune, have a composition that includes heavier elements and ices. The study of exoplanets, planets outside our solar system, has revealed a surprising diversity of planetary systems, challenging our understanding of planet formation and potential habitability.

Galaxies, the massive systems of stars, gas, dust, and dark matter bound together by gravity, are the fundamental building blocks of the universe. Our galaxy, the Milky Way, is a barred spiral galaxy, characterized by its spiral arms and central bar structure. Galaxies come in various shapes and sizes, including elliptical, spiral, and irregular types. Edwin Hubble's classification scheme, known as the Hubble sequence, organizes galaxies based on their morphology. Understanding galaxies' structure and evolution is crucial for studying the universe's history and development.

Dark matter and dark energy are two of the most mysterious and significant concepts in modern astrophysics. Dark matter, which does not emit or absorb light, is thought to make up about 27% of the universe's mass-energy content. Its presence is inferred from its gravitational effects on visible matter, such as the rotation curves of galaxies and the large-scale structure of the universe. Dark energy, an even more enigmatic force, is believed to drive the accelerated expansion of the universe, accounting for about 68% of its mass-energy content. Together, dark matter and dark energy shape the cosmos in ways that are still not fully understood.

The Big Bang theory is the prevailing cosmological model explaining the universe's origin and evolution. According to this theory, the universe began as an incredibly hot and dense singularity about 13.8 billion years ago and has been expanding ever since. The cosmic microwave background radiation, a faint glow left over from the early universe, provides strong evidence for this theory. Studying the Big Bang and the subsequent formation of galaxies, stars, and planets helps us understand the universe's history and future.

Astronomical phenomena, such as eclipses, transits, and occultations, offer valuable opportunities to study celestial objects and their interactions. A solar eclipse occurs when the Moon passes between Earth and the Sun, temporarily blocking the Sun's light. A lunar eclipse happens when Earth passes between the Sun and the Moon, casting a shadow on the Moon. Transits and occultations involve one celestial body passing in front of another, providing insights into their sizes, compositions, and orbits.

Telescope technology has revolutionized our ability to observe and understand the universe. Optical telescopes, which collect visible light, have been the primary tools for centuries. However, other types of telescopes, such as radio, infrared, ultraviolet, X-ray, and gamma-ray telescopes, have expanded our view of the cosmos. Each type of telescope reveals different aspects of celestial objects and phenomena, allowing astronomers to build a more comprehensive picture of the universe.

Astrophysics, the branch of astronomy that deals with the physical properties and processes of celestial objects and phenomena, relies heavily on the principles of physics. Key concepts in astrophysics include gravity, electromagnetism, thermodynamics, and nuclear physics. Understanding these principles helps explain the behavior of stars, black holes, galaxies, and the overall structure and evolution of the universe.

Gravitational waves, ripples in spacetime caused by the acceleration of massive objects, are a groundbreaking discovery in astrophysics. Predicted by Einstein's theory of general relativity, these waves were first directly detected in 2015 by the LIGO and Virgo collaborations. Gravitational waves provide a new way to observe and study cosmic events, such as the merging of black holes and neutron stars, offering insights into the universe's most violent and energetic processes.

Cosmology, the study of the universe's origin, structure, evolution, and eventual fate, is a central field in astronomy. Key cosmological concepts include the Big Bang, cosmic inflation, dark matter, dark energy, and the large-scale structure of the universe. Cosmologists use observations, theoretical models, and simulations to explore fundamental questions about the nature of the cosmos and our place within it.

Astrobiology, the study of life in the universe, seeks to understand the potential for life beyond Earth. This interdisciplinary field combines biology, chemistry, geology, and astronomy to explore the conditions necessary for life and the possibility of habitable environments on other planets and moons. The search for extraterrestrial life involves studying extreme environments on Earth, detecting biosignatures on exoplanets, and exploring the potential habitability of bodies within our solar system, such as Mars and the icy moons of Jupiter and Saturn.

Tools and Techniques in Astrophysics

Telescopes have revolutionized our understanding of the universe, enabling us to observe celestial objects with unprecedented clarity. Optical telescopes, which gather and focus visible light, come in two main types: refracting telescopes that use lenses, and reflecting telescopes that use mirrors. The Hubble Space Telescope, a reflecting telescope, has provided some of the most detailed images of distant galaxies, nebulae, and other astronomical phenomena. Ground-based telescopes, like the Keck Observatory's twin 10-meter mirrors, are located in remote, high-altitude areas to minimize atmospheric distortion and light pollution.

Radio telescopes, another critical tool, detect radio waves emitted by celestial objects. These telescopes can observe phenomena invisible to optical telescopes, such as the cosmic microwave background radiation left over from the Big Bang. The Arecibo Observatory in Puerto Rico, with its 305-meter dish, was once the largest radio telescope in the world, though it collapsed in 2020. The Square Kilometre Array (SKA), currently under development, promises to be the largest and most sensitive radio telescope array, allowing astronomers to probe deeper into the universe's mysteries.

Infrared telescopes observe the infrared part of the electromagnetic spectrum, which lies beyond visible light. These telescopes can see through dust clouds that obscure visible light, revealing star-forming regions and the cores of galaxies. The Spitzer Space Telescope, which operated from 2003 to 2020, provided crucial data on planetary formation and the structure of our galaxy. The James Webb Space Telescope, set to launch soon, will further our understanding of the early universe and the formation of stars and planets.

Ultraviolet (UV) telescopes, like the Hubble Space Telescope, also observe in the UV spectrum, providing insights into the high-energy processes occurring in stars and galaxies. X-ray telescopes, such as the Chandra X-ray Observatory, detect X-rays emitted by hot gases in the universe. These emissions often come from regions around black holes, neutron stars, and supernova remnants, helping astronomers study the universe's most energetic events. Gamma-ray telescopes, like the Fermi Gamma-ray Space Telescope, observe the highest-energy photons, providing data on gamma-ray bursts, pulsars, and active galactic nuclei.

Spectroscopy, the study of the interaction between matter and electromagnetic radiation, is a fundamental technique in astrophysics. By analyzing the light from celestial objects, astronomers can determine their composition, temperature, density, mass, distance, luminosity, and relative motion. Spectrographs, instruments that disperse light into its component wavelengths, are used with telescopes to obtain spectra. The Doppler effect, a change in the frequency of light due to the motion of the source or observer, helps astronomers measure the velocity of stars and galaxies, revealing the expansion of the universe and the presence of exoplanets.

Photometry, the measurement of the intensity of light from an astronomical object, allows astronomers to study variable stars, supernovae, and eclipsing binaries. Light curves, graphs of light intensity over time, provide information on the periodicity and nature of these variations. The Kepler Space Telescope used photometry to detect thousands of exoplanets by observing the tiny dips in starlight as planets transit in front of their host stars.

Astrometry, the precise measurement of the positions and movements of celestial objects, is crucial for mapping the universe and understanding the dynamics of our galaxy. Gaia, a space observatory launched by the European Space Agency, is creating the most accurate 3D map of the Milky Way, measuring the positions, distances, and motions of over a billion stars. This data is vital for studying the structure, formation, and evolution of our galaxy.

Computational astrophysics uses advanced computer simulations to model complex astronomical phenomena, such as galaxy formation, star evolution, and the behavior of black holes. These simulations allow scientists to test theories and interpret observational data. High-performance computing facilities, like NASA's Pleiades supercomputer, are essential for running these large-scale simulations.

Gravitational wave astronomy, a relatively new field, has opened a new window on the universe. Gravitational waves, ripples in spacetime caused by the acceleration of massive objects, were first directly detected in 2015 by the Laser Interferometer Gravitational-Wave Observatory (LIGO). These waves provide information about events such as the mergers of black holes and neutron stars, which cannot be observed with traditional electromagnetic telescopes.

Astrobiology, the study of life in the universe, combines techniques from various scientific disciplines to search for signs of life beyond Earth. Instruments on spacecraft, such as the Mars rovers, analyze soil and rock samples for organic compounds and other potential biosignatures. Space telescopes, like the upcoming James Webb Space Telescope, will examine the atmospheres of exoplanets for gases like oxygen and methane, which could indicate biological activity.

Robotic space missions, such as the Voyager, Cassini, and New Horizons missions, have provided detailed data on the outer planets and their moons. These missions use a variety of instruments, including cameras, spectrometers, magnetometers, and particle detectors, to study the composition, atmosphere, and magnetic fields of these distant worlds. spaceflight, while less common in astrophysics, plays a vital role in some areas of research. The Hubble Space Telescope required several servicing missions by Space Shuttle astronauts to repair and upgrade its instruments. The International Space Station (ISS) hosts experiments in microgravity, which can provide insights into fundamental physical processes that affect astronomical observations.

Citizen science projects, where the public contributes to scientific research, have become increasingly important in astrophysics. Projects like Galaxy Zoo, where volunteers classify galaxies in images from the Sloan Digital Sky Survey, help astronomers analyze vast amounts of data. These efforts not only advance scientific knowledge but also engage and educate the public about astronomy.

Collaborations and international partnerships are essential for large-scale astronomical projects. The

Event Horizon Telescope, a global network of radio telescopes, captured the first image of a black hole in 2019. This achievement required the efforts of hundreds of scientists and institutions worldwide. Similarly, the Large Synoptic Survey Telescope (LSST), set to begin operations soon, is a collaborative effort involving multiple countries and organizations to conduct a comprehensive survey of the night sky.

The future of astrophysics promises even more advanced tools and techniques. Space missions like the European Space Agency's Athena X-ray Observatory and NASA's WFIRST will provide new insights into the universe. Ground-based projects like the Extremely Large Telescope (ELT) and the Thirty Meter Telescope (TMT) will offer unprecedented resolution and sensitivity for studying distant galaxies, exoplanets, and other phenomena.

How to Use This Book

Embarking on a journey through the cosmos with this book as your guide will be both enlightening and exhilarating. It is crafted to be a comprehensive resource, catering to both beginners and those with a bit more experience in the realm of astronomy. The structure is carefully designed to build your knowledge progressively while allowing flexibility in your learning path. Let's delve into the best ways to navigate and utilize this book to maximize your understanding and enjoyment.

Begin by familiarizing yourself with the table of contents. This will give you a broad overview of the topics covered and help you identify areas that

particularly pique your interest or where you may
need a stronger foundation. The chapters are
arranged in a logical order, starting with fundamental
concepts and gradually moving into more advanced
topics. However, the book is also designed to be a
flexible tool. Feel free to jump to chapters that address
your immediate questions or interests. Each chapter is
fairly self-contained, providing a clear and thorough
exploration of its topic.

Starting with the basics is crucial, especially if you are
new to astronomy. The initial chapters cover essential
concepts such as the scale of the universe, the types of
celestial bodies, and basic observational techniques.
These foundational topics will equip you with the
necessary vocabulary and understanding to tackle
more complex subjects later on. As you read, take
notes on key points and terms that are new to you.
This will help reinforce your learning and provide a
handy reference as you progress through the book.

When you encounter particularly challenging sections,
don't be discouraged. Astronomy can be a complex
field, but breaking down difficult concepts into
smaller, manageable parts can make them more
digestible. Revisit these sections after a break or after
reading related chapters that might provide additional
context. Sometimes, understanding a concept requires
a bit of patience and multiple readings. Use the
glossary at the end of the book to clarify any terms
you find confusing.

Engage actively with the material by performing the
suggested exercises and observations. Practical
experience is invaluable in astronomy. If the book
suggests observing the moon's phases or identifying

constellations, take the time to do so. These activities will not only reinforce your theoretical knowledge but also enhance your observational skills. Keeping a journal of your observations can be particularly beneficial. Note the date, time, and conditions of your observations, as well as any interesting details you notice. Over time, you'll build a personal record of your astronomical experiences.

Utilize the diagrams, charts, and images included in the book. Visual aids are a powerful tool in understanding spatial and conceptual relationships in astronomy. Pay close attention to these illustrations, as they often convey information more effectively than text alone. For example, a diagram of the solar system can help you grasp the relative sizes and distances of planets more quickly than a written description. When studying star charts, try to match them with the night sky, either through actual observation or using a planetarium app.

The book includes a variety of sidebars and boxed texts that provide additional insights or interesting anecdotes related to the main content. These sections are designed to enrich your understanding and spark further curiosity. Don't skip them—they often contain fascinating tidbits that can deepen your appreciation of the subject. For instance, a sidebar might explain the mythology behind a constellation or provide a brief biography of a pioneering astronomer.

Discussion questions at the end of each chapter are there to provoke deeper thinking and reinforce your learning. Take the time to ponder these questions and write down your answers. If possible, discuss them with a friend or fellow astronomy enthusiast.

Explaining concepts to others is a powerful way to solidify your own understanding. Additionally, these questions can serve as a springboard for further research and exploration.

Given the vast scope of astronomy, you might find certain topics particularly captivating. When you do, use the references and recommended reading lists provided in the book to delve deeper. Astronomy is a field with a rich history of literature and research, and there's always more to learn. Whether it's a specific celestial phenomenon, a historical figure in astronomy, or the latest discoveries in astrophysics, following your interests will make your study more enjoyable and rewarding.

Community and collaboration are integral to the field of astronomy. Consider joining a local astronomy club or online forum. Sharing your experiences, asking questions, and learning from others can greatly enhance your understanding and passion for the subject. The book might suggest specific groups or forums where beginners are welcome. Engaging with a community can also provide practical tips on equipment, observing locations, and upcoming astronomical events.

Remember that astronomy is a dynamic field. New discoveries and advancements are made regularly. Stay updated by following astronomy news sources, subscribing to relevant magazines, or attending public lectures and events. This book provides a solid foundation, but continuing your education through current resources will keep your knowledge fresh and relevant. Many chapters include suggestions for further reading and resources, which can guide you in staying informed about the latest developments.

Lastly, enjoy the journey. Astronomy is not just about accumulating knowledge; it's about experiencing the wonder of the universe. Take the time to marvel at the night sky, appreciate the beauty of celestial objects, and reflect on our place in the cosmos. Whether you're gazing at the moon through a telescope or reading about distant galaxies, let your curiosity and sense of wonder guide you. This book is a gateway to a lifelong exploration of the universe.

Chapter 2

The Universe An Overview

The Scale of the Universe

The concept of the scale of the universe is one that challenges the boundaries of human imagination. To truly grasp it, one must embark on a mental journey that stretches from the infinitesimal to the infinite. The universe is a vast expanse, filled with a mind-boggling array of objects and phenomena, each existing on a scale so grand that it can be difficult to comprehend. This chapter aims to provide a lens through which you can start to understand the sheer enormity of the cosmos.

Imagine standing on a beach, looking out at the ocean. The horizon seems endless, and yet the ocean is just a tiny fraction of our planet. Now, expand that thought to the Earth itself, which is a small part of our solar system. The solar system, in turn, is just one of many in the Milky Way galaxy. The Milky Way is one of billions of galaxies in the observable universe. Each step outward reveals a larger and more complex structure, highlighting the vastness of the cosmos.

Starting with the smallest scales, we have the realm of subatomic particles. Atoms, the building blocks of matter, are incredibly small, with a diameter of about 0.1 nanometers. Within each atom, protons and neutrons are found in the nucleus, which is about 100,000 times smaller than the atom itself. Electrons orbit this nucleus at distances that are relatively vast compared to the nucleus's size. This structure is akin to a miniature solar system, with the nucleus as the sun and electrons as planets.

Moving up in scale, we reach the level of molecules and cells. Molecules, formed by atoms bonding together, can be as small as a few nanometers. Cells, the basic units of life, range from about 10 micrometers to 100 micrometers in diameter. These scales are still incredibly small, but they form the basis for all life on Earth. The complexity and diversity of life forms begin at this microscopic level.

On the scale of human experience, we measure distances in meters and kilometers. The average height of a person is about 1.7 meters. A marathon spans 42 kilometers. These distances are familiar to us, and they form our immediate perception of the world. Yet, even within our terrestrial bounds, there are vast differences in scale. Mount Everest, the tallest mountain on Earth, stands 8,848 meters above sea level, while the Mariana Trench plunges about 11,000 meters below the ocean's surface.

Expanding our view to the planetary scale, Earth has a diameter of about 12,742 kilometers. It orbits the Sun at an average distance of about 150 million kilometers, a distance known as an astronomical unit (AU). The vastness of our solar system becomes apparent when we consider that Neptune, the outermost planet, orbits the Sun at about 30 AU, or 4.5 billion kilometers. The Oort Cloud, a distant region of icy bodies, extends up to 100,000 AU from the Sun, marking the boundary of the Sun's gravitational influence.

The next leap in scale takes us to the interstellar neighborhood. Our solar system resides in the Milky Way galaxy, a barred spiral galaxy with a diameter of about 100,000 light-years. A light-year, the distance light travels in one year, is roughly 9.46 trillion kilometers. The nearest star to our solar system, Proxima Centauri, is about 4.24 light-years away. This distance is so vast that even the fastest spacecraft would take tens of thousands of years to reach it.

Within the Milky Way, there are an estimated 100 to 400 billion stars, each potentially hosting planetary systems. Our galaxy is just one of many in the Local Group, a collection of about 54 galaxies spread across 10 million light-years. The Andromeda Galaxy, the closest spiral galaxy to the Milky Way, is about 2.537 million light-years away. These distances are difficult to fathom, yet they are just the beginning of the cosmic scale.

Beyond the Local Group lies the Virgo Supercluster, a massive collection of galaxies spanning about 110 million light-years. Our Local Group is situated on the outskirts of this supercluster, which itself is part of an even larger structure known as the Laniakea Supercluster. Laniakea encompasses roughly 100,000 galaxies and spans over 500 million light-years. These superclusters are the largest known structures in the universe, bound together by gravity and forming a vast cosmic web.

On the grandest scale, we consider the observable universe. The observable universe is a sphere with a radius of about 46.5 billion light-years, centered on the observer. This radius corresponds to the age of the universe, approximately 13.8 billion years, multiplied by the speed of light, taking into account the expansion of the universe. Within this volume, there are an estimated 2 trillion galaxies, each containing hundreds of billions of stars.

The scale of the universe can be overwhelming, but it also evokes a sense of wonder and curiosity. The vast distances and enormous structures challenge our understanding and inspire us to explore further. Each new discovery, whether it's a distant exoplanet or a faint galaxy at the edge of the observable universe, adds another piece to the cosmic puzzle.

To appreciate the scale of the universe, it can be helpful to use analogies and models. For example, if the Sun were the size of a basketball, Earth would be a small pea about 25 meters away, and Proxima Centauri would be another basketball over 6,700 kilometers away. Such models can provide a tangible sense of the vast distances involved.

Astronomy tools, such as telescopes and space probes, have expanded our ability to observe and measure the universe. Telescopes, from the humble backyard variety to the sophisticated Hubble Space Telescope, allow us to see distant stars and galaxies. Space probes, like Voyager and New Horizons, provide close-up views of distant planets and their moons. These tools are essential for exploring the universe and understanding its scale.

The scale of the universe also has profound implications for our place within it. The Earth is a tiny speck in a vast cosmic ocean, and our solar system is but one of countless others. This perspective can be humbling, but it also highlights the preciousness of our planet and the uniqueness of life as we know it.

Observing the Cosmos Telescopes and Space Missions

Observing the cosmos is a pursuit that has captivated humanity for centuries. From the early days of stargazing to the sophisticated space missions of today, our quest to understand the universe has driven the development of increasingly advanced tools and techniques. Telescopes and space missions have been at the forefront of this endeavor, allowing us to peer deeper into space and uncover the mysteries of the cosmos.

Telescopes, the fundamental instruments of astronomy, have evolved remarkably since their invention in the early 17th century. The first telescopes, simple refracting models, used lenses to gather and focus light. Galileo Galilei was among the first to use such a telescope for astronomical purposes. His observations of the moons of Jupiter, the phases of Venus, and the rugged surface of the Moon revolutionized our understanding of the solar system.

Modern telescopes come in various types, each designed for specific purposes. Refracting telescopes, like those used by Galileo, are still popular among amateur astronomers for their simplicity and sharp image quality. However, larger and more sophisticated instruments are needed to explore the deeper reaches of space. Reflecting telescopes, which use mirrors instead of lenses to gather light, have become the standard for professional astronomy. The primary mirror of a reflecting telescope can be much larger than a lens, allowing it to collect more light and observe fainter objects.

One of the most famous reflecting telescopes is the Hubble Space Telescope (HST). Launched in 1990, the HST orbits above Earth's atmosphere, providing an unobstructed view of the universe. Its 2.4-meter primary mirror and suite of scientific instruments have produced some of the most detailed and awe-inspiring images of distant galaxies, nebulae, and star clusters. The HST has also contributed to significant discoveries, such as the accelerated expansion of the universe, which led to the concept of dark energy.

Another revolutionary instrument is the James Webb Space Telescope (JWST), set to be the successor to the HST. With a 6.5-meter primary mirror and advanced infrared capabilities, the JWST will peer even further into the cosmos, studying the formation of stars and galaxies in the early universe. Its ability to observe in the infrared spectrum will also allow it to see through dust clouds that obscure many regions of space when viewed in visible light.

Ground-based telescopes have also seen significant advancements. The Very Large Telescope (VLT) in Chile, operated by the European Southern Observatory, consists of four 8.2-meter reflecting telescopes that can be used individually or combined to form an interferometer. This setup allows the VLT to achieve incredibly high-resolution observations, essential for studying distant celestial objects.

Amateur astronomers, too, have access to powerful telescopes that can reveal the wonders of the night sky. Advances in technology have made high-quality refracting and reflecting telescopes more affordable and accessible. Digital imaging and computer-controlled mounts have further enhanced the capabilities of amateur astronomers, enabling them to capture detailed images of planets, galaxies, and nebulae from their backyards.

Beyond telescopes, space missions have played a crucial role in our exploration of the cosmos. Robotic probes and satellites have ventured to the far reaches of the solar system and beyond, sending back invaluable data and images. The Voyager missions, launched in 1977, are among the most famous. Voyager 1 and Voyager 2 have traveled beyond the outer planets, providing detailed images and scientific measurements of Jupiter, Saturn, Uranus, and Neptune. Voyager 1 has even entered interstellar space, becoming the most distant human-made object from Earth.

The Mars rovers, including Spirit, Opportunity, Curiosity, and Perseverance, have transformed our understanding of the Red Planet. These robotic explorers have analyzed Martian soil and rocks, searched for signs of past water activity, and studied the planet's climate and geology. The Perseverance rover, equipped with a suite of advanced instruments, is currently searching for signs of ancient microbial life and collecting samples for future return to Earth.

Space missions have not been limited to our solar system. The Kepler Space Telescope, launched in 2009, revolutionized the search for exoplanets—

planets orbiting stars outside our solar system. Kepler discovered thousands of exoplanets by observing the tiny dips in starlight caused by planets passing in front of their host stars. This mission revealed that planets are common in the galaxy, with many potentially habitable worlds among them.

The Transiting Exoplanet Survey Satellite (TESS), launched in 2018, continues the search for exoplanets, focusing on nearby bright stars. TESS has already identified numerous exoplanet candidates, some of which may be suitable for detailed study by future missions.

Space missions have also explored the outer reaches of the solar system and beyond. The New Horizons mission, launched in 2006, performed a historic flyby of Pluto in 2015, revealing a surprisingly complex and geologically active world. New Horizons then continued its journey into the Kuiper Belt, providing the first close-up images of a Kuiper Belt object, Arrokoth, in 2019.

The study of the cosmos is not limited to visible light. Observatories and space missions operating in other wavelengths, such as radio, infrared, X-ray, and gamma-ray, have uncovered phenomena invisible to optical telescopes. The Chandra X-ray Observatory and the Fermi Gamma-ray Space Telescope, for example, have provided insights into high-energy processes in the universe, such as black holes, neutron stars, and gamma-ray bursts.

The future of observing the cosmos promises even more exciting discoveries. Upcoming missions, such as the European Space Agency's Euclid and the Wide

Field Infrared Survey Telescope (WFIRST), aim to study dark energy and the expansion of the universe. The Square Kilometre Array (SKA), an international effort to build the world's largest radio telescope, will probe the early universe and search for signs of extraterrestrial intelligence.

For beginners looking to observe the cosmos, starting with a small, affordable telescope or even a pair of binoculars can be immensely rewarding. Learning to identify constellations and prominent celestial objects is a good first step. Many resources, including star charts, smartphone apps, and online forums, can help novice astronomers get started.

Patience and practice are essential when learning to use a telescope. Start with bright objects like the Moon, planets, and nearby star clusters. Gradually, as you become more comfortable with your equipment, you can attempt to observe fainter objects like distant galaxies and nebulae. Keeping a log of your observations, noting the date, time, and conditions, can help track your progress and refine your skills.

Joining an astronomy club or participating in public observing events can also enhance your experience. These gatherings provide opportunities to learn from more experienced astronomers, share observations, and gain access to larger telescopes.

The Electromagnetic Spectrum: Seeing the Invisible

The electromagnetic spectrum encompasses all types of electromagnetic radiation, from the shortest gamma rays to the longest radio waves. This spectrum is fundamental to our understanding of the universe, revealing phenomena invisible to the naked eye. Each type of radiation provides a unique window into the cosmos, allowing us to investigate a wide range of celestial objects and processes.

Visible light, the small portion of the spectrum that human eyes can detect, has long been the basis of astronomy. Ancient civilizations relied on visible light to observe the night sky, mapping stars and constellations. However, visible light represents only a tiny fraction of the electromagnetic spectrum. To fully comprehend the universe, astronomers must utilize the entire spectrum, each segment offering distinct insights.

Radio waves, with the longest wavelengths and lowest frequencies, were the first type of non-visible radiation used in astronomy. In the 1930s, Karl Jansky discovered radio waves emanating from the Milky Way, marking the birth of radio astronomy. Radio waves can penetrate dust clouds that obscure visible light, revealing hidden regions of the universe. They are particularly useful for studying cold, dark objects like interstellar gas clouds and the cosmic microwave background—radiation leftover from the Big Bang.

One of the most significant achievements in radio astronomy is the discovery of pulsars—rapidly rotating neutron stars emitting beams of radio waves. Jocelyn Bell Burnell first detected these mysterious objects in 1967. Pulsars provide crucial information about the life cycle of stars and the extreme conditions within neutron stars. Radio telescopes, such as the

Arecibo Observatory and the Very Large Array (VLA), have been instrumental in these discoveries.

Microwaves, with slightly shorter wavelengths than radio waves, also offer valuable insights. The cosmic microwave background (CMB), discovered in 1965 by Arno Penzias and Robert Wilson, is the faint afterglow of the Big Bang. Studying the CMB has allowed scientists to determine the age, composition, and expansion rate of the universe. The Planck satellite, launched in 2009, provided the most detailed map of the CMB, revealing subtle temperature fluctuations that correspond to the seeds of galaxies.

Infrared radiation, with wavelengths longer than visible light, is emitted by warm objects. Infrared astronomy began in earnest in the 1960s with the advent of infrared detectors. This type of radiation is particularly useful for observing star-forming regions, where dust clouds obscure visible light. Infrared telescopes, such as the Spitzer Space Telescope and the Herschel Space Observatory, have unveiled the birthplaces of stars and the structure of galaxies.

The James Webb Space Telescope (JWST), set to launch soon, will be a powerful tool for infrared astronomy. Its large mirror and advanced instruments will allow it to peer through dust clouds and study the formation of stars and planets in unprecedented detail. The JWST's ability to observe the infrared spectrum will also enable it to investigate the atmospheres of exoplanets, searching for signs of habitability and potentially life.

Visible light, while limited in its scope, remains essential for astronomy. Optical telescopes have evolved significantly since Galileo's time, with modern instruments capable of capturing stunningly detailed images of celestial objects. The Hubble Space Telescope, launched in 1990, has revolutionized our understanding of the universe through its observations in visible and ultraviolet light. Hubble's images of distant galaxies, nebulae, and star clusters have captivated the public and provided invaluable data for astronomers.

Ultraviolet (UV) radiation, with shorter wavelengths than visible light, is emitted by hot, young stars and energetic processes. Observing in the ultraviolet spectrum allows astronomers to study the formation and evolution of stars, as well as the behavior of active galaxies and quasars. The Galaxy Evolution Explorer (GALEX) and the Hubble Space Telescope have made significant contributions to ultraviolet astronomy, revealing the dynamic processes at play in the universe.

X-rays, with even shorter wavelengths and higher energies, are produced by some of the most extreme environments in the universe, such as supernova remnants, black holes, and neutron stars. X-ray astronomy began in the 1960s with the launch of the first X-ray satellites. The Chandra X-ray Observatory and the XMM-Newton satellite have provided detailed images and spectra of X-ray sources, helping to uncover the physics of high-energy processes and the structure of galaxy clusters.

Gamma rays, the highest-energy form of electromagnetic radiation, are produced by the most violent events in the cosmos, such as gamma-ray bursts, supernovae, and the behavior of matter near black holes. Gamma-ray astronomy faces significant challenges due to the Earth's atmosphere absorbing these rays. However, space-based observatories like the Fermi Gamma-ray Space Telescope have detected and studied these energetic phenomena, offering insights into the most powerful explosions in the universe.

Understanding the electromagnetic spectrum is crucial for interpreting astronomical data. Each type of radiation provides different information, and combining observations across the spectrum allows for a more comprehensive view of celestial objects and phenomena. For example, studying a galaxy in radio, infrared, visible, ultraviolet, X-ray, and gamma-ray wavelengths can reveal its structure, star formation, chemical composition, and energetic processes.

For beginners interested in exploring the electromagnetic spectrum, starting with visible light observations is a practical first step. A small telescope or even binoculars can reveal a wealth of details about the Moon, planets, and brighter deep-sky objects. As you gain experience and resources, you can explore other wavelengths through online databases and observatories that provide access to radio, infrared, and other types of data.

Many amateur astronomers contribute to the field by participating in citizen science projects, where they can analyze data from professional observatories and help identify new objects or phenomena. Projects like Galaxy Zoo and Zooniverse offer opportunities to engage with data across the electromagnetic spectrum and make meaningful contributions to astronomy.

The electromagnetic spectrum serves as a bridge between the known and the unknown, allowing us to see beyond the limits of our natural vision. By exploring each segment of the spectrum, astronomers can piece together the complex puzzle of the universe. From the cold depths of interstellar space to the fiery hearts of stars, the electromagnetic spectrum reveals the diverse and dynamic nature of the cosmos.

The Expanding Universe and Hubble's Law

In the early 20th century, our understanding of the universe was profoundly transformed by the work of Edwin Hubble. Before Hubble's discoveries, the prevailing view was that the universe was static and unchanging. However, Hubble's observations of distant galaxies led to a groundbreaking realization: the universe is expanding. This expansion is best described by Hubble's Law, which fundamentally altered our perception of the cosmos and laid the groundwork for modern cosmology.

Hubble's journey began with the observation of "nebulae"—cloud-like objects in the night sky. At the time, their true nature was debated. Were these nebulae part of our Milky Way galaxy, or were they separate galaxies far beyond our own? Using the 100-inch Hooker Telescope at Mount Wilson Observatory, Hubble meticulously studied these objects. He identified individual stars within the Andromeda "nebula" and measured their brightness. Among these stars were Cepheid variables, whose intrinsic luminosity could be determined from their pulsation periods. By comparing their observed brightness to their intrinsic brightness, Hubble calculated the distance to Andromeda, conclusively demonstrating that it was a separate galaxy, far outside the Milky Way.

This discovery was monumental, but Hubble's most profound contribution came when he examined the light from these distant galaxies. He noticed a redshift in their spectral lines, indicating that the light from these galaxies was stretched to longer wavelengths. This redshift suggested that the galaxies were moving away from us. To quantify this, Hubble plotted the velocities of galaxies against their distances. The result was a linear relationship: the farther a galaxy is from us, the faster it is receding. This correlation became known as Hubble's Law, mathematically expressed as v=Ho×d$v=Ho×d$, where vv is the galaxy's velocity, dd is its distance, and HoHo is the Hubble constant.

Hubble's Law implies that the universe is expanding uniformly, a concept that was revolutionary at the time. This expansion means that space itself is stretching, carrying galaxies along with it. To visualize this, imagine a balloon with dots representing galaxies on its surface. As the balloon inflates, the dots move away from each other, not because they are traveling through space, but because the space between them is expanding. This analogy, though simplified, helps illustrate the nature of cosmic expansion.

The discovery of the expanding universe had profound implications for cosmology. One of the most significant was the realization that the universe had a beginning—a moment when all matter and energy were concentrated in an incredibly dense state. This idea was the foundation of the Big Bang theory, which posits that the universe began as a singularity approximately 13.8 billion years ago and has been expanding ever since.

Subsequent observations have refined our understanding of Hubble's Law and the Hubble constant. Determining the exact value of the Hubble constant has been a challenging endeavor, with various methods yielding slightly different results. One approach involves observing Cepheid variables and Type Ia supernovae in distant galaxies. These "standard candles" have known luminosities, allowing astronomers to calculate distances accurately. Another method uses the cosmic microwave background (CMB)—the afterglow of the Big Bang—to infer the expansion rate of the early universe. While these methods have improved precision, slight discrepancies remain, sparking ongoing research and debate.

Hubble's Law also suggests that the universe's expansion is not slowing down, but rather accelerating. This surprising discovery was made in the late 1990s by two independent teams studying distant supernovae. They found that these supernovae were dimmer than expected, indicating that the galaxies hosting them were farther away than predicted by a decelerating universe model. This acceleration implies the existence of a mysterious force called dark energy, which counteracts gravity and drives the accelerated expansion. Understanding dark energy remains one of the most significant challenges in modern cosmology.

The implications of an expanding universe extend beyond cosmology to our understanding of fundamental physics. For instance, the concept of an expanding universe led to the development of the cosmological principle, which states that the universe is homogeneous and isotropic on large scales. This principle underpins many cosmological models and theories, including the Lambda Cold Dark Matter (ΛCDM) model, which describes the universe's composition and evolution.

Moreover, the expanding universe framework has profound philosophical and existential implications. It suggests that the universe is dynamic and ever-changing, with a history that can be traced back to a singular origin. This perspective challenges static views of the cosmos and invites us to consider our place in an evolving universe. The notion that the universe had a beginning also raises questions about its ultimate fate. Will the expansion continue indefinitely, leading to a cold, dark, and empty universe? Or will gravitational forces eventually halt and reverse the expansion, resulting in a "Big Crunch"?

For beginners interested in exploring these concepts, a good starting point is to familiarize oneself with the basics of observational astronomy. Understanding how astronomers measure distances and velocities of celestial objects is crucial. Observations of redshifts, for example, can be made using amateur spectroscopes, which split light into its component colors, revealing shifts that indicate motion. Additionally, studying the history of astronomical discoveries provides context for how our understanding of the universe has evolved.

Engaging with citizen science projects can also be an excellent way to delve into cosmology. Projects like Galaxy Zoo invite participants to classify galaxies based on their shapes, contributing to large-scale studies of galaxy formation and evolution. These projects often provide educational resources and community forums where beginners can learn from and collaborate with experienced astronomers.

Dark Matter and Dark Energy

Far beyond the familiar stars and galaxies lies a realm of the universe dominated by two enigmatic substances: dark matter and dark energy. Though invisible and elusive, these components are crucial to understanding the cosmos' structure and fate. Dark matter and dark energy together constitute about 95% of the universe, leaving a mere 5% for the ordinary matter that makes up stars, planets, and everything we can see. Their discovery and study have revolutionized cosmology, providing insights that challenge and expand our understanding of the universe.

Astronomers first inferred the existence of dark matter in the 1930s when Swiss astrophysicist Fritz Zwicky was studying the Coma Cluster, a massive collection of galaxies. Zwicky observed that the galaxies within the cluster were moving too quickly to be held together by the visible matter alone. He proposed the presence of an unseen substance, which he termed "dark matter," to account for the extra gravitational pull needed to keep the cluster intact. Decades later, American astronomer Vera Rubin confirmed the existence of dark matter by observing the rotational speeds of galaxies. She found that stars in the outer regions of galaxies were orbiting faster than expected, suggesting that dark matter was providing the additional gravitational force required to sustain these high velocities.

Dark matter does not emit, absorb, or reflect light, making it invisible to electromagnetic observations. However, its presence can be inferred through its gravitational effects on visible matter, radiation, and the large-scale structure of the universe. One of the most compelling pieces of evidence for dark matter comes from gravitational lensing, where light from distant objects is bent by the gravity of intervening dark matter, creating distorted and magnified images of those objects. The Bullet Cluster, a collision between two galaxy clusters, provides a striking example: while the visible matter of the clusters interacts and slows down, the dark matter passes through without significant interaction, as revealed by the gravitational lensing patterns.

Dark matter's nature remains one of the greatest mysteries in modern physics. It might consist of weakly interacting massive particles (WIMPs), axions, or other exotic particles not yet detected by current experiments. Physicists around the world are conducting numerous experiments to directly detect dark matter particles, using highly sensitive detectors placed deep underground to shield them from cosmic rays and other sources of interference. While these experiments have not yet identified dark matter particles, they have significantly constrained the properties and interaction strengths that such particles might have.

Dark energy, on the other hand, was discovered much more recently, in the late 1990s. Two independent teams of astronomers, studying distant Type Ia supernovae, found that these exploding stars were dimmer than expected. This dimness indicated that the universe's expansion was not slowing down due to gravity, as previously thought, but was instead accelerating. This unexpected discovery suggested the presence of a mysterious force driving this acceleration, which was named dark energy.

Dark energy is even more enigmatic than dark matter. It is thought to be a property of space itself, uniformly filling the universe and exerting a repulsive force that counteracts gravity on cosmic scales. The leading explanation for dark energy is the cosmological constant, first introduced by Albert Einstein as a term in his equations of general relativity. Einstein originally included the cosmological constant to achieve a static universe model, but later abandoned it when the universe was found to be expanding. However, the accelerating expansion discovered by the supernova studies has revived the cosmological constant as a possible explanation for dark energy.

Another intriguing possibility is that dark energy represents a new dynamic field, sometimes called quintessence, which evolves over time. Unlike the cosmological constant, which has a constant energy density, quintessence could vary in strength throughout the history of the universe. Scientists are also exploring alternative theories of gravity that might account for the effects attributed to dark energy without invoking a separate component. These theories propose modifications to general relativity on large scales, potentially eliminating the need for dark energy.

The presence of dark matter and dark energy significantly impacts the universe's fate. In a universe dominated by dark energy, the expansion will continue to accelerate indefinitely, leading to a scenario known as the "Big Freeze" or "Heat Death." In this distant future, galaxies will move farther apart, stars will burn out, and the universe will become increasingly cold and dark. Alternatively, if dark energy's properties change over time, other outcomes are possible, such as the "Big Rip," where the accelerating expansion eventually tears galaxies, stars, and even atoms apart.

For beginners interested in exploring these concepts, a foundational understanding of physics and astronomy is beneficial. Key concepts include gravity, electromagnetism, and the behavior of light. Observational astronomy, particularly the study of galaxy rotations, supernovae, and gravitational lensing, provides critical insights into the evidence for dark matter and dark energy.

Engaging with the scientific community through lectures, online courses, and citizen science projects can also be invaluable. Many universities and research institutions offer free or low-cost courses on cosmology and astrophysics, providing a solid grounding in these subjects. Citizen science projects like Galaxy Zoo, where participants classify galaxies, can offer hands-on experience and a deeper appreciation of the observational techniques used to study the universe.

Chapter 3

The Birth of the Universe The Big Bang Theory

The Origins of the Big Bang Theory

The origins of the Big Bang Theory trace back to the early 20th century, a period marked by revolutionary advancements in our understanding of the universe. Before this era, the prevailing notion among scientists was the idea of a static, eternal universe. This perspective began to shift with the contributions of Einstein, Hubble, and other pioneering astronomers and physicists who laid the groundwork for what would become the most widely accepted explanation for the origin and evolution of the cosmos.

Albert Einstein's general theory of relativity, published in 1915, fundamentally changed our understanding of gravity and space-time. Einstein's equations suggested that the universe could not be static; it had to be either expanding or contracting. However, Einstein himself was uncomfortable with this implication and introduced a cosmological constant, a term that allowed for a static universe by counteracting gravitational collapse. This adjustment was later deemed unnecessary and even mistaken, but it played a crucial role in the development of cosmological theories.

Around the same time, astronomers were making groundbreaking observations that challenged the notion of a static universe. In the 1920s, Edwin

Hubble, using the Hooker Telescope at Mount Wilson Observatory, observed that distant galaxies were moving away from us in all directions. By measuring the redshift of light from these galaxies, Hubble discovered a linear relationship between their distances and velocities, now known as Hubble's Law. This observation provided compelling evidence that the universe was expanding, a finding that would become a cornerstone of the Big Bang Theory.

Georges Lemaître, a Belgian priest and physicist, was one of the first to propose that the universe began from a "primeval atom" or "cosmic egg" that exploded, leading to its continual expansion. Lemaître's 1927 paper on this idea, initially overlooked, gained recognition after Hubble's observations confirmed the expanding universe. Lemaître's hypothesis suggested that the universe had a distinct beginning, challenging the long-held belief in an eternal, unchanging cosmos.

The term "Big Bang" itself was coined somewhat dismissively by British astronomer Fred Hoyle during a 1949 radio broadcast. Hoyle, who favored a competing steady-state theory, intended the term to highlight what he saw as the absurdity of the idea. However, the name stuck and eventually became the accepted label for the theory describing the universe's origin from an extremely hot and dense state.

The Big Bang Theory gained further support through advancements in theoretical physics and observational astronomy. One of the most significant predictions of the theory was the existence of cosmic microwave background radiation (CMB), the faint afterglow of the initial explosion. In 1964, Arno Penzias and Robert Wilson, working at Bell Labs, accidentally discovered this radiation. Their measurements matched the predictions of the Big Bang Theory, providing strong evidence for a hot, dense origin of the universe. Penzias and Wilson were awarded the Nobel Prize in Physics in 1978 for their discovery.

The development of particle physics also played a crucial role in refining the Big Bang Theory. The theory posits that in the first moments after the Big Bang, the universe was in an extremely hot and dense state, with temperatures and energies so high that the fundamental forces of nature were unified. As the universe expanded and cooled, these forces separated in a process known as symmetry breaking. Particle accelerators, like the Large Hadron Collider, have allowed physicists to recreate conditions similar to those just after the Big Bang, providing insights into the behavior of fundamental particles and forces.

The Big Bang Theory also explains the observed abundance of light elements in the universe. According to the theory, during the first few minutes after the Big Bang, conditions were suitable for nuclear reactions to occur, leading to the formation of hydrogen, helium, and a small amount of lithium. This process, known as Big Bang nucleosynthesis, predicted the relative abundances of these elements with remarkable accuracy, matching observations in the universe today.

Despite its successes, the Big Bang Theory has faced challenges and refinements over the years. One of the most significant issues was the horizon problem, which questioned how regions of the universe that were never in causal contact could have the same temperature and properties. The theory of cosmic inflation, proposed by Alan Guth in the early 1980s, addressed this by suggesting a period of rapid exponential expansion in the first fractions of a second after the Big Bang. Inflationary theory not only resolved the horizon problem but also explained the observed large-scale structure of the universe.

Modern cosmology has continued to evolve, incorporating observations from powerful telescopes and satellite missions. The Wilkinson Microwave Anisotropy Probe (WMAP) and the Planck satellite have provided detailed maps of the CMB, allowing scientists to measure the universe's age, composition, and rate of expansion with unprecedented precision. These observations have confirmed many of the predictions of the Big Bang Theory while also revealing the presence of dark matter and dark energy, mysterious components that dominate the universe's mass-energy content.

The Big Bang Theory's implications extend beyond scientific discoveries, influencing our philosophical and existential understanding of the universe. It suggests that the cosmos has a finite history, beginning with an initial singularity and evolving over billions of years. This perspective challenges the notion of an eternal, unchanging universe and invites us to consider the dynamic, ever-changing nature of reality. It raises profound questions about the origins of space and time, the nature of existence, and the ultimate fate of the universe.

For beginners eager to delve into cosmology, understanding the Big Bang Theory involves exploring a range of interconnected concepts and observations.

Grasping the fundamentals of general relativity, the behavior of light, and the properties of elementary particles is essential. Engaging with popular science books, documentaries, and online lectures can provide accessible introductions to these complex topics. Participating in amateur astronomy clubs and observing celestial phenomena can also foster a deeper appreciation of the universe's vastness and the scientific endeavors to understand it.

Evidence Supporting the Big Bang

The universe's inception is one of the most profound questions in cosmology, and the Big Bang Theory stands as the leading explanation. This theory proposes that the universe began from an extremely hot and dense state approximately 13.8 billion years ago and has been expanding ever since. Over the decades, a wealth of evidence has accumulated to support the Big Bang Theory, drawn from diverse fields such as astronomy, particle physics, and observational cosmology. This chapter delves into the critical pieces of evidence that underpin this monumental scientific theory.

One of the most compelling lines of evidence for the Big Bang Theory comes from the observation of the universe's expansion, first noted by Edwin Hubble in the 1920s. Hubble discovered that galaxies are moving away from us in all directions, and the farther away a galaxy is, the faster it appears to be receding. This relationship, now known as Hubble's Law, suggests that the universe is expanding. Hubble's observations provided the first concrete evidence that the universe had a beginning and was not static as previously thought. The linear relationship between distance and velocity of galaxies fits perfectly with the predictions of an expanding universe originating from a single point.

Supporting Hubble's findings, modern measurements of the cosmic expansion rate, known as the Hubble constant, have been refined using various methods. Observations of supernovae, particularly Type Ia supernovae, have been instrumental in this regard. These stellar explosions serve as "standard candles" due to their consistent intrinsic brightness, allowing astronomers to measure distances to galaxies accurately. The data collected from these supernovae confirm that the universe's expansion is not just continuing but accelerating, a discovery that earned the 2011 Nobel Prize in Physics.

The cosmic microwave background (CMB) radiation provides another robust piece of evidence for the Big Bang Theory. The CMB is the afterglow of the hot, dense state of the early universe, now cooled to just 2.7 degrees above absolute zero. Discovered accidentally by Arno Penzias and Robert Wilson in 1964, this faint microwave radiation permeates the entire universe. Its uniformity and slight variations in temperature align with predictions made by the Big Bang Theory. Detailed measurements of the CMB, particularly from the Wilkinson Microwave Anisotropy Probe (WMAP) and the Planck satellite, have provided a precise snapshot of the universe at about 380,000 years old, revealing information about its composition, age, and development.

The Big Bang Theory also accounts for the observed abundances of light elements such as hydrogen, helium, and lithium. In the first few minutes after the Big Bang, conditions were right for nuclear reactions to occur, a process known as Big Bang nucleosynthesis. This process predicts specific ratios of these light elements, which have been observed and measured in the universe. For instance, approximately 75% of the universe's normal matter is hydrogen, and about 25% is helium by mass, with trace amounts of lithium. These proportions closely match the theoretical predictions, providing strong evidence for the theory.

Another compelling piece of evidence comes from the large-scale structure of the universe. The distribution of galaxies and clusters of galaxies is not random but shows a web-like structure with vast voids and dense filaments. These patterns are consistent with the growth of cosmic structures from initial small fluctuations in the density of matter, as seen in the CMB. Over billions of years, these fluctuations grew under the influence of gravity to form the galaxies and clusters we observe today. The clustering of galaxies and the cosmic web provide a natural outcome of the Big Bang Theory, particularly when combined with the theory of cosmic inflation, which posits a rapid expansion of the universe fractions of a second after the Big Bang.

Gravitational waves offer another line of evidence. These ripples in space-time, predicted by Einstein's theory of general relativity, were first directly detected by the Laser Interferometer Gravitational-Wave Observatory (LIGO) in 2015. While these detections were the result of black hole mergers, they also open the door to detecting primordial gravitational waves. Such waves, produced during the inflationary period of the early universe, would provide a direct link to the Big Bang. Although primordial gravitational waves have not yet been detected, their discovery would offer profound confirmation of the inflationary Big Bang model.

Further evidence supporting the Big Bang comes from the observation of the universe's thermal history. As the universe expanded and cooled, it underwent several key phase transitions. The decoupling of matter and radiation, resulting in the formation of the CMB, is one such event. Another is the epoch of reionization, when the first stars and galaxies formed and emitted light that reionized the neutral hydrogen. Observations of distant quasars and the absorption lines in their spectra provide insights into this reionization period, helping to map the thermal history of the universe in line with Big Bang predictions.

The study of distant galaxies and the light they emit also supports the Big Bang Theory. The light from these galaxies has traveled for billions of years to reach us, effectively allowing us to look back in time. Observations of these distant objects reveal that galaxies in the early universe were smaller and less structured than those we see today, consistent with the idea that galaxies formed and evolved over billions of years following the Big Bang. The Hubble Space Telescope and other powerful observatories have captured images of galaxies forming and merging, providing a dynamic picture of cosmic evolution that aligns with the Big Bang model.

In addition to these lines of evidence, the theoretical framework of the Big Bang has been bolstered by advances in particle physics. The conditions of the early universe can be replicated, to some extent, in particle accelerators like the Large Hadron Collider (LHC). Experiments at these facilities have verified aspects of particle behavior under extreme conditions, lending credence to models of the early universe. Discoveries such as the Higgs boson also tie into our understanding of fundamental particles and forces, which play a crucial role in the Big Bang Theory.

Despite the overwhelming evidence, the Big Bang Theory is not without its questions and challenges. For instance, the nature of dark matter and dark energy, which constitute about 95% of the universe's mass-energy content, remains one of the biggest mysteries in cosmology. These components are essential to the current understanding of the universe's expansion and structure, yet they are not fully understood. Ongoing research in astrophysics and cosmology aims to uncover the properties of dark matter and dark energy, potentially leading to new insights or modifications to the Big Bang Theory.

Cosmic Microwave Background Radiation

The universe holds many secrets, but few are as revealing as the Cosmic Microwave Background (CMB) radiation. This faint glow of microwave radiation, permeating the entire cosmos, is a remnant of the early universe, providing a snapshot of the universe at a mere 380,000 years old. The discovery and subsequent study of the CMB have revolutionized our understanding of cosmology, offering profound insights into the origin, composition, and evolution of the universe.

In 1964, Arno Penzias and Robert Wilson, two radio astronomers working at Bell Labs, stumbled upon a persistent noise in their radio telescope. Initially puzzled, they ruled out all possible sources of interference, including urban noise, equipment malfunction, and even pigeon droppings on the antenna. What they discovered was nothing short of groundbreaking—a uniform background radiation that filled the sky. Unbeknownst to them, they had found the CMB, the afterglow of the Big Bang. This discovery provided the first direct evidence of the Big Bang Theory, earning Penzias and Wilson the Nobel Prize in Physics in 1978.

The CMB is a relic from a time when the universe was hot, dense, and almost uniformly filled with a glowing plasma of particles. As the universe expanded, it cooled, allowing protons and electrons to combine and form neutral hydrogen atoms. This period, known as recombination, marked the moment when photons could travel freely through space without being scattered by free electrons, leading to the decoupling of matter and radiation. These photons have been traveling through the universe ever since, gradually stretching to longer wavelengths due to the expansion of the universe and cooling to a temperature of about 2.7 Kelvin, just above absolute zero.

The study of the CMB has been greatly enhanced by various satellite missions, notably the Cosmic Background Explorer (COBE), the Wilkinson Microwave Anisotropy Probe (WMAP), and the Planck satellite. These missions have provided increasingly detailed measurements of the CMB, revealing its tiny temperature fluctuations, or anisotropies. These anisotropies are incredibly small, only about one part in 100,000, but they hold the key to understanding the universe's structure and evolution.

COBE, launched in 1989, was the first satellite to measure the CMB with high precision. It confirmed that the CMB has a nearly perfect blackbody spectrum, as predicted by the Big Bang Theory. This means that the radiation is almost uniformly distributed, with a temperature of about 2.725 Kelvin. However, COBE also detected slight variations in temperature, which were crucial for understanding the formation of galaxies and large-scale structures in the universe. These fluctuations indicated regions of slightly different densities in the early universe, which would eventually grow into galaxies and clusters of galaxies under the influence of gravity.

Following COBE, the WMAP mission, launched in 2001, provided even more detailed measurements of the CMB anisotropies. WMAP mapped the entire sky with much higher resolution, allowing cosmologists to refine their models of the universe. The data from WMAP revealed the age of the universe to be about 13.8 billion years and provided precise measurements of its composition. The results showed that the universe is composed of approximately 4.9% ordinary matter, 26.8% dark matter, and 68.3% dark energy. These findings have been crucial in shaping our current understanding of cosmology.

The Planck satellite, launched in 2009, has provided the most detailed map of the CMB to date. Planck's measurements have refined our understanding of the universe's age, composition, and geometry. The data from Planck have confirmed and extended the findings of WMAP, providing even more precise values for the cosmological parameters. One of the significant discoveries from Planck is the confirmation of the standard model of cosmology, the Lambda Cold Dark Matter (Lambda-CDM) model, which describes the universe as being composed of dark energy, dark matter, and ordinary matter, with dark energy driving the accelerated expansion of the universe.

The CMB also provides insights into the early universe's inflationary period, a rapid expansion that occurred fractions of a second after the Big Bang. Inflation theory predicts that tiny quantum fluctuations during this period were stretched to macroscopic scales, seeding the density variations observed in the CMB. These fluctuations eventually grew into the large-scale structures we see today. The detailed study of the CMB's anisotropies has allowed cosmologists to test and refine inflationary models, providing strong evidence for this period of rapid expansion.

Moreover, the polarization of the CMB offers additional information about the early universe. The CMB is partially polarized due to Thomson scattering, which occurred when photons last interacted with free electrons during recombination. The polarization pattern can be divided into two components: E-modes and B-modes. E-modes are primarily produced by density fluctuations, while B-modes can be generated by gravitational waves from inflation. Detecting B-mode polarization would provide direct evidence for inflation, offering a glimpse into the universe's earliest moments. While B-modes have proven elusive, ongoing experiments and observations continue to search for this critical signal.

The CMB also serves as a cosmic ruler, allowing cosmologists to measure the universe's geometry. By studying the angular size of the temperature fluctuations, scientists can determine whether the universe is flat, open, or closed. The data from WMAP and Planck indicate that the universe is remarkably flat, with a geometry that follows the rules of Euclidean geometry on large scales. This finding has profound implications for our understanding of the universe's overall structure and fate.

Furthermore, the CMB provides a way to test fundamental physics under extreme conditions. The early universe was a hot, dense environment where high-energy processes took place, making it an ideal laboratory for studying particle physics and testing theories beyond the Standard Model. For instance, the interactions of dark matter particles, the properties of neutrinos, and the behavior of fundamental forces at high energies can all be probed through the study of the CMB. As our observational techniques improve, we may uncover new physics that challenges or extends our current theories.

The future of CMB research holds exciting possibilities. Next-generation experiments aim to measure the CMB with even greater precision, potentially detecting the elusive B-mode polarization and providing more insights into the inflationary period. These experiments will also refine our understanding of the universe's composition and evolution, offering the potential for new discoveries about dark matter, dark energy, and fundamental physics.

Nucleosynthesis Formation of the First Elements

In the infancy of our universe, mere moments after the cataclysmic event known as the Big Bang, the cosmos was a seething, chaotic expanse of unimaginable heat and density. It was within this primordial furnace that the first elements began to form, a process known as nucleosynthesis. This epoch, often referred to as Big Bang nucleosynthesis (BBN), set the stage for the chemical evolution of the universe, laying down the foundational elements that would eventually coalesce into stars, planets, and ultimately, life itself.

The story of nucleosynthesis begins in the first few minutes following the Big Bang. During this period, the temperature of the universe was billions of degrees, hot enough to facilitate nuclear reactions. As the universe expanded, it cooled rapidly, moving through a series of critical phases that determined the abundance and types of elements formed.

In the immediate aftermath of the Big Bang, the universe was dominated by a soup of fundamental particles: quarks, electrons, neutrinos, and photons. As the universe cooled, quarks combined to form protons and neutrons. This phase, occurring within the first second, was critical because the number of protons relative to neutrons would influence the subsequent formation of elements.

Approximately one minute after the Big Bang, the universe had cooled to a temperature at which protons and neutrons could begin to combine to form the nuclei of the first elements. However, this process was not straightforward. Neutrons are slightly heavier than protons and are prone to decay into protons, electrons, and antineutrinos with a half-life of about ten minutes. The balance between protons and neutrons was therefore in constant flux, governed by the interplay of nuclear reactions and the relentless ticking of neutron decay.

Despite these challenges, the cooling universe eventually reached a temperature conducive to the formation of deuterium, a stable isotope of hydrogen composed of one proton and one neutron. This process, known as deuterium bottleneck, was a pivotal moment in nucleosynthesis. As soon as deuterium nuclei formed, they could quickly fuse with other protons and neutrons to create heavier elements like helium-3, helium-4, and traces of lithium-7.

The production of helium-4 was particularly significant. Helium-4 nuclei, consisting of two protons and two neutrons, were exceptionally stable and thus became the most abundant product of Big Bang nucleosynthesis after hydrogen. By the time the universe was about three minutes old, most of the neutrons had been incorporated into helium-4 nuclei, resulting in a primordial ratio of about 75% hydrogen and 25% helium by mass. This ratio is remarkably consistent across the observable universe, providing strong evidence for the Big Bang model.

While hydrogen and helium were the primary products of Big Bang nucleosynthesis, trace amounts of other light elements were also formed. Lithium-7, for instance, was produced in small quantities, though the observed abundance of lithium in the universe has posed some intriguing puzzles for cosmologists, known as the "lithium problem." This discrepancy between predicted and observed lithium-7 abundances suggests that there may be additional processes at play, either during or after the era of Big Bang nucleosynthesis.

The formation of heavier elements, beyond lithium, required environments with much higher temperatures and densities than those present in the early universe. These conditions would only be found later, within the cores of stars and during supernova explosions. Thus, the first few minutes of the universe set the stage for the formation of the simplest and most abundant elements, while the complexity of the periodic table would unfold over billions of years of stellar evolution.

As we delve deeper into the details of nucleosynthesis, we encounter the concept of nuclear binding energy, which plays a crucial role in the formation and stability of atomic nuclei. Binding energy is the energy required to split a nucleus into its constituent protons and neutrons. Nuclei with higher binding energy per nucleon are more stable. Helium-4, with its high binding energy, is thus more stable than many other light nuclei, explaining its abundance.

The study of nucleosynthesis also sheds light on the conditions of the early universe and the fundamental forces that govern it. The rates of nuclear reactions, the density and temperature of the universe, and the influence of fundamental forces such as gravity and electromagnetism all played a part in determining the outcomes of Big Bang nucleosynthesis. By comparing theoretical predictions with observed abundances of light elements, cosmologists can test and refine their models of the early universe.

One of the triumphs of modern cosmology is the agreement between the predicted abundances of light elements from Big Bang nucleosynthesis and their observed values. This concordance provides compelling evidence for the Big Bang model and the standard cosmological framework. The precise measurements of the cosmic microwave background radiation, the relic radiation from the Big Bang, have further corroborated these findings, allowing scientists to infer the conditions of the early universe with remarkable accuracy.

The journey of nucleosynthesis did not end with the Big Bang. Stellar nucleosynthesis, the process by which elements are formed within stars, took over from where Big Bang nucleosynthesis left off. In the cores of stars, nuclear fusion reactions continue to forge heavier elements, building upon the primordial abundance of hydrogen and helium. Over successive generations of star formation and supernova explosions, elements like carbon, oxygen, and iron were synthesized and dispersed throughout the cosmos, enriching the interstellar medium and paving the way for the formation of planets and life.

The study of nucleosynthesis also intersects with other fields of physics and astronomy. For instance, the properties of neutrinos, elusive particles that played a significant role during Big Bang nucleosynthesis, have implications for particle physics and the search for new physics beyond the Standard Model. Observations of ancient stars, which preserve the chemical signatures of early nucleosynthesis, provide valuable data for testing cosmological theories. Furthermore, the ongoing search for primordial gravitational waves, ripples in spacetime produced in the early universe, could offer new insights into the conditions that prevailed during the era of Big Bang nucleosynthesis.

The Evolution of the Universe Post-Big Bang

The universe, born from the explosive event known as the Big Bang approximately 13.8 billion years ago, embarked on a journey of evolution and transformation that has led to the cosmos we observe today. This grand narrative encompasses a series of epochs, each characterized by significant physical processes and milestones that shaped the structure and composition of the universe. From the formation of the first atoms to the assembly of galaxies and the birth of stars, the evolution of the universe post-Big Bang is a story of complexity emerging from simplicity.

In the immediate aftermath of the Big Bang, the universe was a hot, dense plasma of fundamental particles. During the first few microseconds, quarks and gluons combined to form protons and neutrons in a process driven by the strong nuclear force. As the universe continued to expand and cool, these protons and neutrons began to combine to form the nuclei of the first elements, primarily hydrogen and helium, during a period known as Big Bang nucleosynthesis. This phase, which occurred within the first three minutes, laid the groundwork for the chemical composition of the cosmos.

As the universe expanded further, it entered a period known as the "photon epoch," which lasted for about 380,000 years. During this time, the universe was still too hot for electrons to bind with nuclei to form neutral atoms. Instead, the universe was filled with a plasma of free electrons, protons, and helium nuclei, along with a sea of photons. These photons constantly scattered off the free electrons in a process known as Thomson scattering, making the universe opaque to radiation.

A major turning point occurred when the universe cooled sufficiently to allow electrons to combine with protons and helium nuclei to form neutral atoms, a process called recombination. This event marked the end of the photon epoch and the beginning of the "dark ages," a period when the universe became transparent to radiation. The photons released during recombination have traveled through space nearly unimpeded since that time, forming the cosmic microwave background (CMB) radiation, a faint glow that provides a snapshot of the universe at that early stage.

The dark ages were so named because, with the formation of neutral atoms, the universe lacked any sources of light. This period lasted for several hundred million years until the first stars and galaxies began to form. These first sources of light, called Population III stars, were massive, short-lived, and composed almost entirely of hydrogen and helium. Their formation marked the end of the dark ages and initiated the epoch of reionization.

During the epoch of reionization, the intense ultraviolet radiation from the first stars ionized the surrounding hydrogen gas, gradually reionizing the universe. This process created ionized bubbles around individual stars and galaxies, which eventually overlapped, making the universe transparent to ultraviolet light once again. By the end of reionization, which occurred around one billion years after the Big Bang, most of the hydrogen in the universe was ionized.

The formation of the first stars and galaxies also heralded the beginning of hierarchical structure formation in the universe. Galaxies formed within dark matter halos, regions of higher density that gravitationally attracted baryonic matter. Over time, these galaxies merged and interacted, leading to the formation of larger structures such as galaxy clusters and superclusters. The distribution of these structures forms a cosmic web, a vast network of filaments and voids that trace the underlying dark matter distribution.

As galaxies evolved, so did the stars within them. Stellar nucleosynthesis, the process by which stars produce heavier elements through nuclear fusion, enriched the interstellar medium with elements such as carbon, oxygen, and iron. These elements were dispersed into space through supernova explosions, the violent deaths of massive stars. This enrichment played a crucial role in the formation of subsequent generations of stars and planetary systems, including our own solar system.

The universe's expansion, driven by the mysterious force known as dark energy, continued to accelerate over time. Observations of distant supernovae and the large-scale structure of the cosmos have provided strong evidence for this accelerated expansion. Dark energy, which constitutes about 68% of the total energy density of the universe, remains one of the most profound mysteries in cosmology.

Throughout this cosmic journey, the interplay of gravity, dark matter, dark energy, and baryonic matter has shaped the evolution of the universe. Gravity, the dominant force on large scales, has driven the formation of structures through the collapse of matter into gravitational wells. Dark matter, which makes up about 27% of the universe's energy density, has provided the scaffolding for structure formation. Its presence is inferred from its gravitational effects, though it remains elusive in terms of direct detection.

Dark energy, on the other hand, acts as a repulsive force, counteracting gravity on the largest scales and driving the accelerated expansion of the universe. Its exact nature is unknown, but it is often modeled as a cosmological constant or as a dynamic field that evolves over time. Understanding dark energy is one of the primary goals of modern cosmology, as it holds the key to the ultimate fate of the universe.

The evolution of the universe has also been marked by several key observational discoveries. The discovery of the cosmic microwave background radiation in 1965 provided strong evidence for the Big Bang theory and offered a glimpse into the early universe. Observations of the large-scale structure of the universe, including galaxy surveys and measurements of the cosmic web, have revealed the distribution of matter on vast scales and the role of dark matter in structure formation.

In addition, the study of distant supernovae has provided insights into the universe's expansion history and the nature of dark energy. These observations have led to the development of the standard cosmological model, known as the Lambda Cold Dark Matter (ΛCDM) model, which describes the universe's composition and evolution with remarkable precision.

As we look to the future, several upcoming observational missions and experiments promise to deepen our understanding of the universe's evolution. The James Webb Space Telescope, launched in 2021, is already providing unprecedented views of the early universe, allowing astronomers to study the formation of the first galaxies and stars. Future missions, such as the Euclid satellite and the Vera C. Rubin Observatory, will map the large-scale structure of the universe with greater precision, shedding light on the nature of dark matter and dark energy.

Chapter 4

Galaxies Islands in the Universe

Types of Galaxies Spiral, Elliptical, and Irregular

Galaxies, the vast systems of stars, gas, dust, and dark matter bound together by gravity, come in a variety of shapes and sizes. These cosmic islands are categorized into three primary types: spiral, elliptical, and irregular galaxies. Each type exhibits distinct characteristics and evolutionary histories, contributing to the rich tapestry of the universe. Understanding these types not only provides insight into the life cycle of galaxies but also reveals the underlying processes that govern their formation and development.

Spiral galaxies are among the most striking and well-studied types of galaxies. They are characterized by their flat, disk-like structure with a central bulge and spiral arms that emanate outward. The Milky Way, our home galaxy, is a quintessential example of a spiral galaxy. These galaxies are further classified into two main subcategories: normal spirals and barred spirals.

Normal spiral galaxies, such as the Andromeda Galaxy, feature well-defined spiral arms that wind outward from the central bulge. These arms are sites of active star formation, populated by young, hot, blue stars that illuminate the surrounding gas and dust.

The spiral pattern is maintained by density waves that propagate through the disk, compressing gas and triggering the birth of new stars.

Barred spiral galaxies, on the other hand, possess a distinct bar-shaped structure that extends through the central bulge, with spiral arms originating from the ends of the bar. The presence of the bar is thought to influence the dynamics of the galaxy, funneling gas toward the center and potentially fueling active galactic nuclei. An example of a barred spiral galaxy is the galaxy NGC 1300, where the bar's influence is clearly visible in its structure.

The formation of spiral galaxies is believed to be linked to the rotation of the initial gas cloud from which they formed. As this cloud collapses under gravity, it begins to spin, flattening into a disk. The conservation of angular momentum ensures that the disk continues to rotate, giving rise to the characteristic spiral structure. Interactions with other galaxies, such as minor mergers and tidal forces, can also play a role in shaping and maintaining the spiral pattern.

Elliptical galaxies, in contrast to spirals, are more spheroidal or ellipsoidal in shape and lack the distinct spiral arms. They range from nearly spherical (classified as E0) to highly elongated (classified as E7). These galaxies are typically composed of older, redder stars and have little to no active star formation. The stars in elliptical galaxies exhibit random orbits, giving the galaxy a more uniform and featureless appearance.

The formation of elliptical galaxies is thought to result from the merging of smaller galaxies. When two or more galaxies collide, their stars, gas, and dark matter interact gravitationally, leading to the formation of a more spheroidal structure. During these violent mergers, gas is often driven toward the center, where it can trigger intense bursts of star formation before being depleted or heated to temperatures that prevent further star formation.

Elliptical galaxies can be found in a variety of environments, from isolated regions to the dense cores of galaxy clusters. In clusters, frequent interactions and mergers are more likely, contributing to the prevalence of elliptical galaxies in these regions. The largest known galaxies, such as the giant elliptical galaxy M87 in the Virgo Cluster, are often the result of multiple merging events over cosmic time.

Irregular galaxies, as their name suggests, do not fit neatly into the categories of spiral or elliptical galaxies. They lack a distinct shape and often appear chaotic, with no central bulge or defined spiral arms. These galaxies are typically rich in gas and dust, with active star formation occurring throughout their irregular structure. The Large and Small Magellanic Clouds, which are satellite galaxies of the Milky Way, are classic examples of irregular galaxies.

The irregular appearance of these galaxies can be attributed to several factors. In some cases, irregular galaxies may be the result of gravitational interactions or mergers that have disrupted their original structure. In other instances, they may represent a more primitive stage of galaxy evolution, where the processes that lead to the formation of well-defined structures have not yet fully occurred.

Irregular galaxies are often found in close proximity to larger galaxies, where tidal forces and gravitational interactions can distort their shapes. These interactions can also trigger bursts of star formation, as gas is compressed and funneled into regions where new stars can form. Despite their chaotic appearance, irregular galaxies play a crucial role in the overall picture of galaxy evolution, providing insights into the early stages of galaxy formation and the effects of interactions in shaping galactic structures.

The classification of galaxies into spiral, elliptical, and irregular types was first systematically undertaken by the American astronomer Edwin Hubble in the early 20th century. Hubble's "tuning fork" diagram, which visually represents these classifications, remains a foundational tool in the study of galaxies. While Hubble's scheme has been refined and expanded over the years, it still provides a useful framework for understanding the diversity of galactic forms.

The study of galaxies extends beyond mere classification, as astronomers strive to understand the underlying physical processes that govern their formation and evolution. Advances in observational technology, such as the Hubble Space Telescope and other space-based observatories, have allowed astronomers to peer deeper into the universe and observe galaxies at various stages of their life cycles. These observations have revealed that galaxies are dynamic entities, constantly evolving through interactions, mergers, and internal processes.

For instance, the phenomenon of galactic cannibalism, where larger galaxies merge with and consume smaller ones, has been observed in numerous cases. The Milky Way itself is currently in the process of merging with the Sagittarius Dwarf Elliptical Galaxy, and evidence suggests that it has undergone similar mergers in the past. These interactions contribute to the growth and evolution of galaxies, adding complexity to their structures and influencing their star formation histories.

The role of dark matter in galaxy formation and evolution is another area of active research. Dark matter, which does not emit or absorb light, constitutes about 27% of the universe's mass-energy content and exerts a significant gravitational influence on visible matter. It is thought to provide the scaffolding for galaxy formation, with galaxies forming within dark matter halos. Observations of galaxy rotation curves and gravitational lensing provide indirect evidence for the presence of dark matter, shaping our understanding of its role in the cosmos.

The Milky Way Our Galactic Home

The Milky Way, our galactic home, is a sprawling collection of stars, gas, dust, and dark matter bound together by gravity. This majestic structure, seen as a milky band of light stretching across the night sky, holds countless secrets about the cosmos and our place within it. The Milky Way is not just an astronomical entity; it is a crucial part of our history, our future, and our understanding of the universe.

Our galaxy is a barred spiral galaxy, one of the most common types in the universe. It consists of several key components: the central bulge, the disk, the spiral arms, and the halo. Each of these parts plays a significant role in the galaxy's structure and dynamics.

The central bulge is a dense, spheroidal region located at the heart of the Milky Way. It is populated by old stars, some of which are among the oldest in the galaxy. The bulge also contains a supermassive black hole known as Sagittarius A*, which has a mass equivalent to about four million suns. This black hole is a crucial player in the dynamics of the central region, influencing the orbits of stars and the behavior of gas clouds.

Surrounding the central bulge is the galactic disk, a vast, flattened region that contains the majority of the Milky Way's stars, including our Sun. The disk is about 100,000 light-years in diameter and only about 1,000 light-years thick. It is within this disk that we find the spiral arms, regions of higher density that wind outward from the center. These arms are sites of active star formation, where gas and dust coalesce to form new stars and planetary systems. The Sun resides in a minor arm called the Orion Arm, located about 27,000 light-years from the galactic center.

Beyond the disk lies the halo, a roughly spherical region that extends far beyond the main body of the galaxy. The halo contains older stars, globular clusters, and a significant amount of dark matter. The presence of dark matter is inferred from its gravitational effects on the visible components of the galaxy, and it is believed to make up most of the Milky Way's mass. The halo also hosts the Magellanic

Clouds, two small satellite galaxies that orbit the Milky Way and interact with it gravitationally.

The Milky Way is in constant motion. Stars and gas clouds orbit the galactic center, with speeds varying depending on their distance from the center. The Sun, for example, completes one orbit around the galaxy every 225 to 250 million years. This motion is not uniform, as stars can migrate between different parts of the galaxy over time due to gravitational interactions and other processes. This dynamic nature of the Milky Way is a testament to the complex interplay of forces that shape our galactic home.

The history of the Milky Way is a story of formation and evolution. It is believed to have formed around 13.6 billion years ago from a primordial cloud of gas and dark matter. Over time, this cloud collapsed under its own gravity, leading to the formation of the first stars and star clusters. These early structures merged and interacted, giving rise to the complex structure we see today. The Milky Way has also grown by accreting smaller galaxies, a process that continues to this day. Evidence of past mergers can be seen in the form of stellar streams and other features that trace the remnants of these interactions.

The future of the Milky Way is equally intriguing. One of the most significant events on the horizon is the predicted collision with the Andromeda Galaxy, our nearest large neighbor. This cosmic encounter is expected to occur in about 4.5 billion years. When the two galaxies collide, they will merge to form a new, larger galaxy. While this may sound cataclysmic, the vast distances between stars mean that direct collisions between stars are unlikely. Instead, the gravitational forces will reshape the galaxies, leading to new patterns of star formation and potentially creating a new supermassive black hole at the center of the merged galaxy.

Understanding the Milky Way also means studying the myriad stars and planetary systems within it. Our galaxy is home to a diverse array of stars, from massive, short-lived giants to small, long-lived dwarfs. Stellar evolution, the process by which stars are born, live, and die, is a fundamental aspect of our understanding of the cosmos. The death of stars leads to the creation of white dwarfs, neutron stars, and black holes, each of which plays a role in the galactic ecosystem.

Planetary systems, like our own solar system, are also a key focus of study. The discovery of exoplanets—planets orbiting other stars—has revolutionized our understanding of the Milky Way. These discoveries have revealed that planets are common and diverse, ranging from gas giants larger than Jupiter to rocky worlds similar to Earth. The search for potentially habitable exoplanets is a major area of research, driven by the hope of finding signs of life beyond our solar system.

The Milky Way's interstellar medium, the space between stars filled with gas and dust, is another critical area of study. This medium is the birthplace of stars and planets, and its properties influence the processes of star formation and galactic evolution. The study of the interstellar medium involves understanding the complex interactions between gas, dust, magnetic fields, and cosmic rays.

Our place in the Milky Way offers a unique vantage point for exploring these phenomena. As inhabitants of one of the galaxy's spiral arms, we have the opportunity to observe and study our galactic neighbors. Astronomy and space exploration missions, such as the Hubble Space Telescope and the upcoming James Webb Space Telescope, provide unprecedented views of the Milky Way and beyond. These instruments allow us to peer into the heart of the galaxy, study distant star-forming regions, and uncover the secrets of the interstellar medium.

The cultural and historical significance of the Milky Way cannot be overlooked. Throughout human history, the Milky Way has inspired myths, legends, and scientific inquiry. Ancient civilizations saw it as a river of milk or a path of souls, while early astronomers sought to understand its nature and origin. Today, the Milky Way continues to captivate our imagination and drive our quest for knowledge.

Galaxy Formation and Evolution

The Milky Way, our galactic home, is a sprawling collection of stars, gas, dust, and dark matter bound together by gravity. This majestic structure, seen as a milky band of light stretching across the night sky, holds countless secrets about the cosmos and our place within it. The Milky Way is not just an astronomical entity; it is a crucial part of our history, our future, and our understanding of the universe.

Our galaxy is a barred spiral galaxy, one of the most common types in the universe. It consists of several key components: the central bulge, the disk, the spiral arms, and the halo. Each of these parts plays a significant role in the galaxy's structure and dynamics.

The central bulge is a dense, spheroidal region located at the heart of the Milky Way. It is populated by old stars, some of which are among the oldest in the galaxy. The bulge also contains a supermassive black hole known as Sagittarius A*, which has a mass equivalent to about four million suns. This black hole is a crucial player in the dynamics of the central region, influencing the orbits of stars and the behavior of gas clouds.

Surrounding the central bulge is the galactic disk, a vast, flattened region that contains the majority of the Milky Way's stars, including our Sun. The disk is about 100,000 light-years in diameter and only about 1,000 light-years thick. It is within this disk that we find the spiral arms, regions of higher density that wind outward from the center. These arms are sites of active star formation, where gas and dust coalesce to form new stars and planetary systems. The Sun resides in a minor arm called the Orion Arm, located about 27,000 light-years from the galactic center.

Beyond the disk lies the halo, a roughly spherical region that extends far beyond the main body of the galaxy. The halo contains older stars, globular clusters, and a significant amount of dark matter. The presence of dark matter is inferred from its gravitational effects on the visible components of the galaxy, and it is believed to make up most of the Milky Way's mass. The halo also hosts the Magellanic Clouds, two small satellite galaxies that orbit the Milky Way and interact with it gravitationally.

The Milky Way is in constant motion. Stars and gas clouds orbit the galactic center, with speeds varying depending on their distance from the center. The Sun, for example, completes one orbit around the galaxy every 225 to 250 million years. This motion is not uniform, as stars can migrate between different parts of the galaxy over time due to gravitational interactions and other processes. This dynamic nature of the Milky Way is a testament to the complex interplay of forces that shape our galactic home.

The history of the Milky Way is a story of formation and evolution. It is believed to have formed around 13.6 billion years ago from a primordial cloud of gas and dark matter. Over time, this cloud collapsed under its own gravity, leading to the formation of the first stars and star clusters. These early structures merged and interacted, giving rise to the complex structure we see today. The Milky Way has also grown by accreting smaller galaxies, a process that continues to this day. Evidence of past mergers can be seen in the form of stellar streams and other features that trace the remnants of these interactions.

The future of the Milky Way is equally intriguing. One of the most significant events on the horizon is the predicted collision with the Andromeda Galaxy, our nearest large neighbor. This cosmic encounter is expected to occur in about 4.5 billion years. When the two galaxies collide, they will merge to form a new, larger galaxy. While this may sound cataclysmic, the vast distances between stars mean that direct collisions between stars are unlikely. Instead, the gravitational forces will reshape the galaxies, leading to new patterns of star formation and potentially creating a new supermassive black hole at the center of the merged galaxy.

Understanding the Milky Way also means studying the myriad stars and planetary systems within it. Our galaxy is home to a diverse array of stars, from massive, short-lived giants to small, long-lived dwarfs. Stellar evolution, the process by which stars are born, live, and die, is a fundamental aspect of our understanding of the cosmos. The death of stars leads to the creation of white dwarfs, neutron stars, and black holes, each of which plays a role in the galactic ecosystem.

Planetary systems, like our own solar system, are also a key focus of study. The discovery of exoplanets—planets orbiting other stars—has revolutionized our understanding of the Milky Way. These discoveries have revealed that planets are common and diverse, ranging from gas giants larger than Jupiter to rocky worlds similar to Earth. The search for potentially habitable exoplanets is a major area of research, driven by the hope of finding signs of life beyond our solar system.

The Milky Way's interstellar medium, the space between stars filled with gas and dust, is another critical area of study. This medium is the birthplace of stars and planets, and its properties influence the processes of star formation and galactic evolution. The study of the interstellar medium involves understanding the complex interactions between gas, dust, magnetic fields, and cosmic rays.

Our place in the Milky Way offers a unique vantage point for exploring these phenomena. As inhabitants of one of the galaxy's spiral arms, we have the opportunity to observe and study our galactic neighbors. Astronomy and space exploration missions, such as the Hubble Space Telescope and the upcoming James Webb Space Telescope, provide unprecedented views of the Milky Way and beyond. These instruments allow us to peer into the heart of the galaxy, study distant star-forming regions, and uncover the secrets of the interstellar medium.

The cultural and historical significance of the Milky Way cannot be overlooked. Throughout human history, the Milky Way has inspired myths, legends, and scientific inquiry. Ancient civilizations saw it as a river of milk or a path of souls, while early astronomers sought to understand its nature and origin. Today, the Milky Way continues to captivate our imagination and drive our quest for knowledge.

Interactions and Collisions Between Galaxies

When we look up at the sky on a clear night, we witness a still and serene universe. However, the cosmos is far from tranquil. Galaxies, those vast islands of stars, gas, and dark matter, are constantly interacting and sometimes even colliding with one another. These galactic encounters play a crucial role in the evolution of galaxies, including our own Milky Way. Understanding these interactions and collisions provides insights into the dynamic and ever-changing nature of the universe.

Galaxy interactions can range from gentle gravitational tugs to violent mergers. These interactions are governed by the fundamental force of gravity, which dictates the motions and distortions of galaxies as they pass near each other. When two galaxies come close, their mutual gravitational attraction can pull stars, gas, and dust out of their original orbits, creating spectacular tidal tails and bridges of material. These tidal features are often the first visible signs of galactic interaction.

One of the most well-known examples of interacting galaxies is the Antennae Galaxies, located about 45 million light-years away in the constellation Corvus. These two spiral galaxies are in the process of merging, and their interaction has created long, sweeping tidal tails that resemble an insect's antennae. The collision has triggered intense star formation, as the gas clouds within the galaxies are compressed and collapse to form new stars. This burst of star formation is often seen in interacting galaxies and is a direct consequence of the gravitational forces at play.

Galactic collisions can take millions or even billions of years to complete. During a collision, the stars within the galaxies generally do not collide directly due to the vast distances between them. Instead, the gravitational forces cause the galaxies to distort and merge, while their gas clouds can collide and compress, leading to the formation of new stars. This process can result in the creation of new structures, such as elliptical galaxies, which are often the remnants of spiral galaxy mergers.

The Milky Way is not immune to such interactions. In fact, it has already experienced several minor mergers with smaller galaxies, and it is currently interacting with the Sagittarius Dwarf Spheroidal Galaxy. This small galaxy is being torn apart by the Milky Way's gravity, and its stars are being assimilated into our galaxy. These minor mergers contribute to the growth and evolution of the Milky Way, adding new stars and dark matter to its halo.

The most significant galactic collision in the Milky Way's future is its predicted merger with the Andromeda Galaxy, the nearest spiral galaxy to us. This event is expected to occur in about 4.5 billion years. Both galaxies are currently on a collision course, and when they eventually merge, they will form a new, larger galaxy, often referred to as "Milkomeda" or "Milkdromeda." This merger will dramatically reshape both galaxies, creating new star formation regions and potentially leading to the formation of a new supermassive black hole at the center of the merged galaxy.

While the idea of galaxies colliding may seem catastrophic, these events are a natural part of the universe's evolution. They play a crucial role in shaping galaxies and driving their development. During a merger, the intense gravitational interactions can funnel gas towards the center of the galaxies, fueling the growth of supermassive black holes. These black holes can, in turn, influence the formation of stars and the distribution of gas and dust within the galaxy.

Galaxy interactions and mergers also provide a unique laboratory for studying the properties of dark matter.

Dark matter is an invisible substance that makes up about 27% of the universe's mass-energy content. It does not emit, absorb, or reflect light, making it detectable only through its gravitational effects. When galaxies collide, their dark matter halos pass through each other relatively unaffected, while the visible matter interacts more strongly. By studying these interactions, astronomers can infer the distribution and behavior of dark matter, providing crucial clues about its nature.

One of the most famous examples of using galaxy collisions to study dark matter is the Bullet Cluster. This system consists of two colliding galaxy clusters, with their respective dark matter halos and hot gas clouds. Observations of the Bullet Cluster have shown that the dark matter halos have passed through each other, while the hot gas has interacted and slowed down. This separation of dark matter and visible matter provides strong evidence for the existence of dark matter and its role in the universe.

The study of galaxy interactions and collisions also helps us understand the formation and evolution of galaxy clusters. Galaxy clusters are the largest gravitationally bound structures in the universe, containing hundreds or even thousands of galaxies. These clusters form through the merging of smaller groups of galaxies and individual galaxies over billions of years. The interactions within clusters can lead to the stripping of gas from galaxies, quenching star formation, and transforming spiral galaxies into elliptical ones.

Observations of galaxy clusters, such as the Coma Cluster and the Virgo Cluster, reveal a complex web of interactions and mergers. These clusters provide a snapshot of the dynamic processes shaping galaxies on large scales. By studying the distribution of galaxies, gas, and dark matter within clusters, astronomers can gain insights into the history of these structures and the forces driving their evolution.

The advent of advanced telescopes and observational techniques has revolutionized our understanding of galaxy interactions and collisions. Instruments like the Hubble Space Telescope and the Atacama Large Millimeter/submillimeter Array (ALMA) have provided detailed images and data on interacting galaxies, revealing the intricate structures and processes involved. Computer simulations have also played a crucial role, allowing astronomers to model and predict the outcomes of galactic encounters.

These simulations, such as the Illustris and EAGLE projects, have provided valuable insights into the dynamics of galaxy interactions and the formation of structures within galaxies. By comparing these simulations with observations, astronomers can test their theories and refine their models, leading to a deeper understanding of the universe.

Galaxy interactions and collisions are not just a topic of scientific inquiry; they also capture the imagination of the public. The dramatic images of colliding galaxies, with their tidal tails and starburst regions, are a testament to the beauty and complexity of the cosmos. These events remind us of the dynamic nature of the universe and our place within it.

As we continue to study galaxy interactions and collisions, we are likely to uncover even more surprises and deepen our understanding of the forces shaping the universe. Each new discovery adds to the rich tapestry of knowledge, helping us piece together the history and future of galaxies.

Active Galactic Nuclei and Quasars

Astronomers have long been fascinated by the extraordinary phenomena occurring at the centers of some galaxies, known as Active Galactic Nuclei (AGN). These regions are incredibly luminous, sometimes outshining the rest of the galaxy combined. At the heart of AGN lies a supermassive black hole, millions to billions of times the mass of the Sun, surrounded by an accretion disk of infalling material. As this material spirals into the black hole, it heats up and emits vast amounts of energy across the electromagnetic spectrum. Among the most extreme examples of AGN are quasars, which are the universe's brightest and most energetic objects.

The discovery of quasars dates back to the 1960s when astronomers first identified these mysterious sources of radio waves. Optical observations revealed that these sources had unusual spectra with high redshifts, indicating that they were incredibly distant and, therefore, enormously powerful. The realization that quasars were not stars within our galaxy but rather the luminous cores of distant galaxies revolutionized our understanding of the universe. These objects were named "quasi-stellar objects," or quasars, because of their star-like appearance.

At the core of a quasar is a supermassive black hole consuming material at an astonishing rate. This accretion process releases tremendous energy, making quasars visible across vast cosmic distances. The accretion disk around the black hole glows brilliantly, and powerful jets of particles are often launched from the regions near the black hole at nearly the speed of light. These jets can extend for thousands of light-years, interacting with the intergalactic medium and influencing the evolution of surrounding galaxies.

Understanding the mechanics of AGN and quasars requires delving into the physics of accretion disks and relativistic jets. The accretion disk forms as gas and dust fall toward the black hole, heating up due to friction and gravitational forces. The temperature of the disk can reach millions of degrees, emitting X-rays and ultraviolet radiation. This intense radiation can ionize surrounding gas, creating vast clouds of ionized material that emit characteristic spectral lines.

The process of jet formation is less well understood but is thought to involve the twisting of magnetic fields in the accretion disk. These magnetic fields can channel particles into narrow beams that are propelled away from the black hole at relativistic speeds. As these jets interact with the interstellar and intergalactic medium, they can create enormous lobes of radio emission, which are characteristic of many radio-loud quasars.

AGN are classified based on their observational properties, which can vary depending on the orientation of the accretion disk and jets relative to our line of sight. Seyfert galaxies, for example, have relatively modest luminosities compared to quasars and often exhibit strong emission lines from ionized gas. Radio galaxies, on the other hand, are characterized by their powerful radio jets and lobes. Blazars are a particularly extreme type of AGN where the jet is pointed almost directly at Earth, resulting in rapid variability and high-energy emissions.

One of the key challenges in studying AGN and quasars is understanding their role in the broader context of galaxy evolution. It is now widely accepted that nearly every massive galaxy hosts a supermassive black hole at its center, and the growth of these black holes is closely linked to the development of their host galaxies. During periods of intense accretion, AGN can dramatically influence their surroundings through feedback processes.

AGN feedback can heat and expel gas from the central regions of galaxies, regulating star formation and shaping the galaxy's evolution. This feedback can take the form of powerful winds driven by radiation pressure from the accretion disk or mechanical energy from relativistic jets. Observations of galaxy clusters, such as the Perseus Cluster, reveal vast cavities in the hot intracluster medium created by AGN jets, demonstrating the significant impact these processes can have on their environments.

Quasars also serve as valuable probes of the early universe. Because they are so luminous, they can be observed at great distances, providing a glimpse into the conditions of the universe when it was only a few billion years old. Studying high-redshift quasars helps astronomers understand the formation and growth of the first supermassive black holes, as well as the reionization of the intergalactic medium.

The most distant quasars known are seen at redshifts greater than 7, corresponding to a time when the universe was less than a billion years old. These quasars challenge our understanding of black hole formation, as their existence implies that supermassive black holes grew rapidly in the early universe. Various theories have been proposed to explain this rapid growth, including the formation of massive seed black holes from the collapse of primordial gas clouds or the merging of smaller black holes.

The study of AGN and quasars is not limited to observations. Theoretical models and computer simulations play a crucial role in unraveling the complex physics of these objects. Simulations of accretion disks, jet formation, and AGN feedback help researchers test their theories and predict the behavior of these systems under different conditions. These models must account for a wide range of physical processes, from the microphysics of particle acceleration in jets to the large-scale impact of AGN on galaxy evolution.

Multiwavelength observations are essential for studying AGN and quasars, as these objects emit across the entire electromagnetic spectrum. Radio telescopes, such as the Very Large Array (VLA), can image the jets and lobes of radio-loud AGN, while optical and infrared telescopes, like the Hubble Space Telescope and the James Webb Space Telescope, provide detailed views of the accretion disk and host galaxy. X-ray observatories, such as Chandra and XMM-Newton, probe the high-energy emission from the innermost regions of the accretion disk, and gamma-ray telescopes, like the Fermi Gamma-ray Space Telescope, detect the most energetic photons from AGN jets.

The study of AGN and quasars is a rapidly evolving field, with new discoveries and insights emerging regularly. The next generation of observatories, including the Square Kilometre Array (SKA) and the Extremely Large Telescope (ELT), will provide unprecedented sensitivity and resolution, enabling astronomers to study these objects in greater detail than ever before. These advancements will help answer fundamental questions about the nature of black holes, the mechanisms driving AGN activity, and the role of AGN in galaxy evolution.

Chapter 5

Stars The Building Blocks of Galaxies

Star Formation From Nebulae to Protostars

Gazing up at the night sky, one might wonder how stars, those twinkling points of light, come into being. The journey from vast, diffuse clouds of gas and dust to the brilliant, burning stars is a tale of gravity, turbulence, and time. This process, known as star formation, begins in regions of space known as nebulae. Nebulae are the cradles of stars, places where the conditions are just right for the birth of these luminous objects.

Nebulae come in various forms, but the ones most associated with star formation are called molecular clouds. These clouds are cold and dense, composed primarily of hydrogen molecules, along with helium and trace amounts of heavier elements. The temperatures in these clouds can be as low as a few degrees above absolute zero, making them some of the coldest places in the universe. Despite their frigid temperatures, molecular clouds are bustling with activity, as gravity pulls the gas and dust together, creating regions of higher density.

Within these dense regions, called clumps, the initial stages of star formation begin. Each clump can contain enough material to form multiple stars. The process is often triggered by external forces, such as the shock waves from nearby supernovae or the pressure from stellar winds of massive stars. These forces compress the clumps, causing them to collapse under their own gravity. As the clump collapses, it fragments into smaller cores, each of which can form a star.

As gravity pulls the material inwards, the cores become denser and hotter. The collapsing gas forms a rotating disk around a central concentration of mass. This central mass, known as a protostar, is the nascent stage of a star. The surrounding disk is made up of gas and dust that will eventually accrete onto the protostar or form planets, moons, and other objects in a developing star system.

The journey from a protostar to a fully-fledged star is marked by several key stages. Initially, the protostar is not yet hot enough to initiate nuclear fusion, the process that powers stars. Instead, it shines dimly due to the conversion of gravitational energy into heat and light. This phase can last for millions of years as the protostar continues to gather material from its surrounding disk.

As the protostar accretes more mass, its core temperature and pressure increase. When the core temperature reaches around 10 million degrees Celsius, hydrogen nuclei begin to fuse into helium, releasing vast amounts of energy in the form of light and heat. This marks the birth of a true star. The onset of nuclear fusion halts the gravitational collapse, as the outward pressure from the fusion reactions balances the inward pull of gravity. The star enters the main sequence phase of its life, burning hydrogen in its core and shining steadily.

Star formation does not occur in isolation. The birth of a new star can influence its surroundings in profound ways. The intense radiation from young, massive stars can ionize nearby gas, creating bright emission nebulae. Stellar winds and supernova explosions from massive stars can trigger the formation of new stars by compressing nearby gas clouds. This cycle of star birth and death can shape entire galaxies, driving their evolution over billions of years.

The formation of stars also leads to the creation of stellar clusters. These clusters are groups of stars that form from the same molecular cloud and are bound together by gravity. They can range from small groups of a few dozen stars to massive clusters containing hundreds of thousands of stars. Studying these clusters provides valuable insights into the processes of star formation and the early stages of stellar evolution.

While the basic outline of star formation is well understood, many details remain the subject of active research. For example, the role of magnetic fields in regulating the collapse of molecular clouds and the formation of protostars is an area of ongoing investigation. Magnetic fields can influence the rate at which material accretes onto the protostar and can help launch powerful jets of material that are often observed emanating from young stars.

Another area of interest is the formation of massive stars, which are more challenging to study due to their rarity and the fact that they form quickly and in densely clustered environments. Massive stars have a profound impact on their surroundings, producing strong stellar winds and eventually exploding as supernovae. These processes can trigger the formation of new stars and enrich the interstellar medium with heavy elements, influencing the chemical evolution of galaxies.

Observations across different wavelengths of light are crucial for studying star formation. Radio and millimeter-wave telescopes can peer through the dense dust clouds to reveal the cold gas and dust in molecular clouds. Infrared observations can detect the heat from protostars still embedded in their natal clouds. Optical and ultraviolet telescopes can observe the young stars and the ionized gas in their surroundings. Each wavelength provides a different piece of the puzzle, helping astronomers build a comprehensive picture of the star formation process.

Theoretical models and computer simulations also play a vital role in understanding star formation. These models can simulate the complex interplay of gravity, turbulence, magnetic fields, and radiation within molecular clouds. By comparing the results of simulations with observations, astronomers can test their theories and refine their understanding of the processes involved.

The study of star formation has practical implications as well. Understanding how stars form and evolve is essential for understanding the history and future of our own Sun and solar system. It also has implications for the search for life beyond Earth. The conditions in star-forming regions can influence the formation of planets and the potential for habitable environments. By studying star formation, we can learn more about the origins of the elements that make up planets and life itself.

The Life Cycle of Stars Main Sequence to Supernova

Stars, the dazzling beacons scattered across the night sky, undergo a life cycle as dynamic and varied as life itself. They are born, they live, and they die, each phase marked by profound changes in their structure and appearance. The journey of a star from the main sequence to its dramatic end as a supernova is a tale of nuclear fusion, gravitational forces, and cosmic transformations.

Once a star has formed from the collapse of a molecular cloud and entered the main sequence phase, it spends the majority of its life in a state of relative stability. The main sequence is essentially a period of hydrogen burning, where the star fuses hydrogen into helium in its core. This fusion process releases enormous amounts of energy, which provides the outward pressure needed to balance the inward pull of gravity. This balance, known as hydrostatic equilibrium, keeps the star stable and shining brightly.

The length of time a star spends on the main sequence depends heavily on its mass. Massive stars burn their fuel much more quickly than their smaller counterparts, leading to shorter lifespans. For example, a star like our Sun will remain on the main sequence for about 10 billion years, whereas a more massive star might only last a few million years. In contrast, smaller, cooler stars, known as red dwarfs, can burn for tens to hundreds of billions of years.

As a star exhausts the hydrogen in its core, the balance of forces within it changes. Without the outward push of hydrogen fusion, the core begins to contract under gravity. This contraction increases the core's temperature and pressure until conditions are ripe for the fusion of helium into heavier elements like carbon and oxygen. During this phase, the star swells into a red giant or, in the case of very massive stars, a supergiant. The outer layers expand and cool, giving the star its characteristic red color.

For stars like the Sun, the red giant phase is a relatively peaceful transition. Helium fusion in the core generates sufficient pressure to halt further collapse and stabilize the star for a while. However, this phase is much shorter than the main sequence, lasting only a few hundred million years. Eventually, the star will shed its outer layers, creating a beautiful shell of ionized gas known as a planetary nebula. The remaining core, no longer able to sustain fusion reactions, becomes a white dwarf—a dense, Earth-sized remnant that slowly cools and fades over billions of years.

Massive stars, however, follow a more tumultuous path. After burning through their helium, they have enough mass and pressure to fuse heavier elements in a series of successive stages. Each stage of fusion releases less energy, leading to shorter and shorter lifespans for each phase. The star builds up layers like an onion, with successive shells of fusion surrounding an iron core. Iron fusion is endothermic, meaning it absorbs energy rather than releasing it, which spells disaster for the star.

When the core of a massive star becomes predominantly iron, it can no longer support itself against gravitational collapse. The core rapidly contracts, and the outer layers of the star fall inward. This catastrophic implosion rebounds off the core, creating a shock wave that propels the outer layers into space in a titanic explosion known as a supernova. The energy released in a supernova is so immense that, for a brief period, the star can outshine an entire galaxy.

The remnants of a supernova can take several forms, depending on the mass of the original star. If the core remnant is between about 1.4 and 3 times the mass of the Sun, it will become a neutron star—a city-sized object so dense that a single teaspoon of its material would weigh billions of tons. Neutron stars exhibit extraordinary properties, including incredibly strong magnetic fields and rapid rotation speeds, sometimes observed as pulsars.

For the most massive stars, those with core remnants exceeding three solar masses, the collapse continues until a black hole forms. Black holes are regions of space where gravity is so strong that not even light can escape. They profoundly influence their surroundings, often found at the centers of galaxies or in binary systems where they can accrete material from a companion star, emitting powerful X-rays in the process.

The death of stars, particularly in supernovae, plays a crucial role in the cosmos. These cataclysmic events scatter heavy elements like carbon, oxygen, and iron into space, enriching the interstellar medium with the building blocks for new stars, planets, and even life. In this way, the life cycle of stars is a fundamental process that drives the evolution of galaxies and the chemical complexity of the universe.

In addition to their dramatic ends, the remnants of dead stars continue to impact their environments. Planetary nebulae and supernova remnants can trigger the formation of new stars by compressing nearby gas clouds. The elements forged in the hearts of stars and spread by supernovae become part of the next generation of stars and planets, continuing the cosmic cycle of birth, life, and death.

The study of stars and their life cycles is a vibrant field of astronomy, driven by observations across the electromagnetic spectrum. Telescopes that detect different types of light—radio, infrared, optical, ultraviolet, X-ray, and gamma-ray—provide a comprehensive view of the processes at work. Theoretical models and computer simulations complement these observations, helping astronomers understand the intricate physics involved in stellar evolution.

Despite the progress made, many mysteries remain. For instance, the exact mechanisms behind supernova explosions are still not fully understood. Researchers continue to investigate the role of neutrinos—ghostly particles produced in vast numbers during a supernova—in the explosion process. Similarly, the formation of black holes and neutron stars, particularly in binary systems, presents ongoing challenges and opportunities for discovery.

The life cycle of stars is not just a tale of individual objects but a story of the universe itself. From the peaceful glow of main sequence stars to the violent deaths of supernovae, stars shape the cosmos in profound ways. They are the forges where elements are created, the engines driving galactic evolution, and the sources of the light that illuminates our universe. Understanding their life cycles enriches our knowledge of the natural world and our place within it.

Types of Stars Dwarfs, Giants, and Neutron Stars

Stars, the luminous spheres of plasma scattered across the cosmos, come in a remarkable variety of types, each with its own unique characteristics and life cycles. Understanding the different types of stars—dwarfs, giants, and neutron stars—provides insight into the complex processes that govern their formation, evolution, and ultimate fate.

Dwarf stars form the backbone of the galaxy, being the most numerous type of star. Among them, red dwarfs are the smallest and coolest, with masses ranging from about 0.08 to 0.6 times that of the Sun. These stars have relatively low temperatures, between 2,500 and 4,000 Kelvin, giving them a distinct reddish hue. Red dwarfs burn their hydrogen fuel very slowly, allowing them to shine for tens to hundreds of billions of years, far longer than stars like our Sun. This slow burn results in a stable and prolonged main sequence phase, making red dwarfs some of the oldest stars in the universe. Despite their longevity, red dwarfs are often too dim to be seen with the naked eye from Earth.

Moving up the scale, we encounter the more familiar yellow dwarfs, like our Sun. These stars have masses between 0.6 and 1.4 times that of the Sun and temperatures ranging from 5,000 to 6,000 Kelvin. Yellow dwarfs spend approximately 10 billion years on the main sequence, fusing hydrogen into helium in their cores. This process provides a steady source of energy, resulting in a stable and relatively long-lived star. As yellow dwarfs age and exhaust their hydrogen fuel, they expand into red giants before shedding their outer layers and leaving behind a white dwarf.

White dwarfs are the remnants of medium-sized stars like the Sun. After the red giant phase, these stars expel their outer layers and leave behind a dense, Earth-sized core. White dwarfs no longer undergo fusion reactions; instead, they gradually cool and fade over billions of years. Despite their small size, white dwarfs are incredibly dense, with a single teaspoon of their material weighing several tons. Their intense gravity causes a phenomenon known as electron degeneracy pressure, which prevents further collapse.

In contrast to the relatively modest dwarfs, giant stars are characterized by their substantial size and brightness. Red giants are stars that have exhausted the hydrogen in their cores and have begun burning helium into heavier elements. As the core contracts, the outer layers expand and cool, giving the star a reddish appearance. Red giants can have diameters tens to hundreds of times greater than that of the Sun, and their luminosity can be thousands of times greater. The red giant phase is a relatively brief period in a star's life, lasting only a few hundred million years before the star sheds its outer layers and leaves behind a white dwarf.

In the category of giants, we also find supergiants, which are among the most massive and luminous stars in the universe. Supergiants, both red and blue, have masses ranging from 10 to over 100 times that of the Sun. These stars burn through their nuclear fuel at an astonishing rate, leading to relatively short lifespans of a few million years. Blue supergiants are extremely hot, with surface temperatures exceeding 20,000 Kelvin, while red supergiants are cooler but vastly larger. Supergiants often end their lives in spectacular supernova explosions, leaving behind neutron stars or black holes.

Neutron stars are the remnants of massive stars that have undergone supernova explosions. These stars are incredibly dense, with masses between 1.4 and 3 times that of the Sun but compressed into a sphere only about 20 kilometers in diameter. A neutron star's density is so extreme that protons and electrons combine to form neutrons, giving the star its name. Neutron stars exhibit some of the most extreme physical properties in the universe. They have incredibly strong magnetic fields, often trillions of times stronger than Earth's, and can rotate at astonishing speeds, sometimes hundreds of times per second. These rapidly rotating neutron stars are known as pulsars, emitting beams of electromagnetic radiation that sweep across space like the beams of a lighthouse.

The study of stars also reveals interesting subcategories and special cases within these broad types. For instance, brown dwarfs are often referred to as "failed stars." With masses between the heaviest gas giant planets and the lightest stars (about 13 to 80 times the mass of Jupiter), brown dwarfs are not massive enough to sustain hydrogen fusion in their cores. Instead, they emit faint infrared radiation, making them challenging to detect.

Blue stragglers are another intriguing class of stars. Found in old star clusters where most stars have evolved off the main sequence, blue stragglers appear younger and hotter than their neighbors. Their origin is still a topic of research, with theories suggesting they may form through the merger of two stars or by siphoning material from a companion star.

Variable stars, including Cepheid variables and RR Lyrae stars, provide crucial tools for measuring astronomical distances. These stars pulsate in brightness due to changes in their outer layers. The regularity of their pulsations allows astronomers to determine their intrinsic brightness and, consequently, their distance from Earth. Cepheid variables, in particular, have been instrumental in measuring the scale of the universe.

Stars can also form binary or even multiple star systems, where two or more stars orbit a common center of mass. Such systems can exhibit complex interactions, including mass transfer between stars, which can significantly affect their evolution. In some cases, binary systems can lead to phenomena such as Type Ia supernovae, where a white dwarf accretes material from a companion star until it reaches a critical mass and explodes.

The diversity of stars, from dwarfs to giants to neutron stars, highlights the intricate and varied nature of stellar evolution. Each type of star follows a unique path dictated by its initial mass, composition, and environment. These differences in stellar evolution contribute to the rich tapestry of the universe, driving the chemical enrichment of galaxies and the formation of new stars and planetary systems.

As our understanding of stars continues to grow, so does our appreciation for the complexity and beauty of the cosmos. Observations across the electromagnetic spectrum, from radio waves to gamma rays, combined with advances in theoretical modeling, provide ever-deeper insights into the lives of stars. The study of these celestial objects not only enhances our knowledge of the universe but also offers a glimpse into the processes that have shaped our own solar system and the conditions that make life possible.

Stellar Nucleosynthesis Birth of Elements

Stars, the cosmic forges, play a fundamental role in the creation of elements through a process known as stellar nucleosynthesis. This fascinating phenomenon begins with the simplest of elements and, through complex nuclear reactions, builds up the periodic table we know today. Stellar nucleosynthesis is not just a tale of fusion and fission but a narrative of the universe's evolution, giving birth to the very elements that make up planets, life, and the cosmos itself.

The story starts with hydrogen, the most abundant element in the universe. In the dense, hot cores of stars, hydrogen nuclei, or protons, collide with such force that they overcome their mutual repulsion and fuse together. This fusion forms helium, the second simplest element, in a process known as hydrogen burning or the proton-proton chain. The energy released in this fusion powers the star, producing the light and heat we see and feel. For stars like our Sun, this hydrogen burning is the primary source of energy for billions of years, maintaining a delicate balance between the inward pull of gravity and the outward pressure from nuclear fusion.

As stars exhaust their hydrogen fuel, they enter subsequent phases of nucleosynthesis. In more massive stars, once hydrogen is depleted in the core, the temperature and pressure rise sufficiently to ignite helium. Helium nuclei, or alpha particles, combine to form carbon through the triple-alpha process, which involves the fusion of three helium nuclei. This process not only produces carbon, a fundamental building block of life, but also generates oxygen when additional helium nuclei fuse with carbon. Thus, the cores of aging stars become enriched with these heavier elements.

The lifecycle of a star is intricately tied to its mass. While stars like the Sun eventually shed their outer layers and leave behind a dense core known as a white dwarf, more massive stars continue to forge even heavier elements. As the core contracts and the temperature rises, these stars undergo successive stages of nuclear burning. After helium burning, carbon burning begins, followed by neon, oxygen, and silicon burning. Each stage synthesizes progressively heavier elements, culminating in the formation of iron in the core.

Iron represents a turning point in stellar nucleosynthesis. Unlike lighter elements, whose fusion releases energy, the fusion of iron and heavier elements requires an input of energy. Consequently, once a star's core is predominantly iron, further fusion becomes energetically unfavorable. For massive stars, this marks the end of their life cycle. Unable to support the immense gravitational pressure, the core collapses, leading to a catastrophic supernova explosion. This explosion not only disperses the elements formed in the star's core into space but also provides the necessary conditions for the creation of elements heavier than iron.

Supernovae are responsible for some of the most fascinating elements in the universe. The intense heat and pressure of a supernova enable rapid neutron capture processes, known as the r-process, which can produce elements much heavier than iron, such as gold, platinum, and uranium. These elements are scattered across the galaxy, seeding future generations of stars and planetary systems with the building blocks of planets and life.

Interestingly, not all heavy elements are formed in massive stars and supernovae. Some are created in the relatively peaceful environments of asymptotic giant branch (AGB) stars. These stars, which are in a late stage of evolution, experience periods of intense nucleosynthesis in their outer layers. Through slow neutron capture processes, known as the s-process, elements such as strontium, barium, and lead are formed. These elements are then dredged up to the surface and eventually expelled into space through stellar winds, contributing to the chemical enrichment of the interstellar medium.

The cycle of stellar nucleosynthesis is a continuous one. The elements produced in stars are ejected into space through supernova explosions, stellar winds, and other processes, enriching the interstellar medium with heavy elements. These enriched clouds of gas and dust then collapse to form new stars and planetary systems. Our own solar system, for example, is thought to have formed from the remnants of previous generations of stars, incorporating elements forged in distant stellar furnaces.

One of the most compelling pieces of evidence for stellar nucleosynthesis comes from the study of stellar spectra. By analyzing the light emitted by stars, astronomers can determine their chemical composition. These observations reveal that older stars, which formed earlier in the universe's history, have lower abundances of heavy elements compared to younger stars. This gradient in chemical composition, known as metallicity, supports the idea that successive generations of stars have progressively enriched the universe with heavy elements.

Stellar nucleosynthesis also has profound implications for our understanding of the universe. The distribution of elements in the cosmos, from the hydrogen and helium formed in the Big Bang to the heavy elements produced in stars, provides crucial insights into the processes that have shaped the evolution of galaxies and the cosmos as a whole. For example, the presence of certain isotopes in ancient meteorites can reveal information about the nucleosynthetic processes that occurred in the early solar system.

Moreover, the study of stellar nucleosynthesis has practical applications closer to home. Understanding the origins of elements and their distribution in the galaxy can inform the search for habitable exoplanets and the potential for life beyond Earth. The elements necessary for life, such as carbon, nitrogen, and oxygen, are products of stellar nucleosynthesis, and their presence in exoplanetary systems is a key factor in assessing their habitability.

The intricate dance of particles in the heart of stars, driven by the forces of gravity and nuclear fusion, is a story of creation and transformation. Stellar nucleosynthesis is a testament to the dynamic and interconnected nature of the universe, where the life and death of stars are inextricably linked to the formation of elements and the evolution of galaxies. As we continue to explore the cosmos, the study of stellar nucleosynthesis will undoubtedly remain a cornerstone of our quest to understand the origins and destiny of the elements that make up the universe.

The Hertzsprung-Russell Diagram

The Hertzsprung-Russell Diagram (HR Diagram) is a cornerstone of modern astrophysics, providing a graphical representation of the relationship between the luminosity, temperature, and spectral types of stars. This diagram, named after Danish astronomer

Ejnar Hertzsprung and American astronomer Henry Norris Russell, has revolutionized our understanding of stellar evolution and the life cycles of stars.

Stars, like people, have life stages. They are born, mature, and eventually die. The HR Diagram is akin to a family photo album, capturing stars at various stages of their lives and offering a snapshot of their evolutionary paths. The fundamental axes of the HR Diagram are luminosity, plotted on the vertical axis, and surface temperature, plotted on the horizontal axis. Luminosity increases upward, while temperature decreases from left to right, which is somewhat counterintuitive but traditional in the field.

At the heart of the HR Diagram lies the main sequence, a diagonal band stretching from the top left (hot, luminous stars) to the bottom right (cool, dim stars). About 90% of the stars, including our Sun, reside on the main sequence during the majority of their lifetimes, fusing hydrogen into helium in their cores. The position of a star on the main sequence is determined primarily by its mass: more massive stars are hotter and more luminous, while less massive stars are cooler and less luminous.

Above the main sequence, in the upper right quadrant of the HR Diagram, lie the red giants and supergiants. These stars have exhausted the hydrogen in their cores and have expanded and cooled as they burn helium or heavier elements in shells around the core. Despite their cooler surface temperatures, red giants and supergiants are incredibly luminous due to their enormous sizes. Betelgeuse in the constellation Orion is a famous example of a red supergiant.

In contrast, the lower left quadrant of the HR Diagram is home to white dwarfs, the remnants of stars like our Sun that have shed their outer layers and ceased nuclear fusion. These objects are incredibly hot but very dim due to their small sizes. White dwarfs are essentially the exposed cores of former stars, slowly cooling and fading over billions of years.

The HR Diagram also reveals the existence of different stellar populations and the processes that govern their evolution. For instance, stars that are more massive than the Sun evolve off the main sequence more quickly, becoming red giants or supergiants in a relatively short time. These massive stars end their lives in dramatic supernova explosions, often leaving behind neutron stars or black holes. Conversely, less massive stars, such as red dwarfs, burn their fuel much more slowly and can remain on the main sequence for tens to hundreds of billions of years.

To fully appreciate the HR Diagram's utility, it's essential to consider how astronomers use it to study star clusters. Star clusters, groups of stars that formed from the same molecular cloud and are roughly the same age, provide a unique laboratory for testing theories of stellar evolution. By plotting the positions of stars in a cluster on the HR Diagram, astronomers can determine the cluster's age. This technique relies on identifying the "turn-off point," where stars are just beginning to leave the main sequence. The position of this turn-off point indicates the age of the cluster, as stars of different masses leave the main sequence at different times.

The HR Diagram also plays a crucial role in understanding the chemical composition of stars. Spectroscopy, the study of how stars absorb and emit light, allows astronomers to determine the elements present in a star's atmosphere. When combined with the star's position on the HR Diagram, this information can reveal details about the star's past and its place in the broader context of galactic evolution. For example, older stars tend to have lower metallicities (fewer elements heavier than helium) than younger stars, reflecting the gradual enrichment of the galaxy with heavy elements over time.

Another fascinating aspect of the HR Diagram is its ability to illustrate the effects of binary star systems on stellar evolution. In close binary systems, where two stars orbit each other in close proximity, mass transfer can occur from one star to the other. This process can significantly alter the evolutionary paths of both stars, leading to phenomena such as blue stragglers—main sequence stars that appear younger and more massive than they should be for their cluster's age.

The HR Diagram is also instrumental in the study of variable stars, which are stars that exhibit changes in brightness over time. Different types of variable stars occupy distinct regions of the HR Diagram. For instance, Cepheid variables, which pulsate with regular periods, are found in a region known as the instability strip, crossing the main sequence. By studying these variables, astronomers can gain insights into the internal structures and processes of stars.

Despite its apparent simplicity, the HR Diagram encapsulates a wealth of information about the physical properties and evolutionary histories of stars. It serves as a roadmap, guiding astronomers in their quest to unravel the mysteries of the cosmos. Each point on the diagram represents a star, a beacon of light in the vastness of space, with its own unique story woven into the fabric of the universe.

Understanding the HR Diagram also underscores the interconnectedness of the universe. The life cycle of stars, from their formation in stellar nurseries to their demise in supernovae or as white dwarfs, is a cosmic dance that has profound implications for the formation of planets and the development of life. The elements forged in the hearts of stars are the building blocks of everything we see around us, from the iron in our blood to the silicon in our electronics.

The HR Diagram, with its elegant simplicity and profound depth, remains one of the most powerful tools in the astronomer's toolkit. It bridges the gap between observation and theory, allowing us to piece together the life stories of stars and, by extension, the history of the universe itself. As we continue to refine our models and gather more data from telescopes and space missions, the HR Diagram will undoubtedly continue to illuminate our path, guiding us toward a deeper understanding of the cosmos and our place within it.

www.ingramcontent.com/pod-product-compliance
Lightning Source LLC
Chambersburg PA
CBHW071209130726
47998CB00002B/678